Classification of
Lipschitz Mappings

PURE AND APPLIED MATHEMATICS

A Program of Monographs, Textbooks, and Lecture Notes

MONOGRAPHS AND TEXTBOOKS IN PURE AND APPLIED MATHEMATICS

Recent Titles

Robert Carlson, A Concrete Introduction to Real Analysis (2006)

John Dauns and Yiqiang Zhou, Classes of Modules (2006)

N. K. Govil, H. N. Mhaskar, Ram N. Mohapatra, Zuhair Nashed, and J. Szabados, Frontiers in Interpolation and Approximation (2006)

Luca Lorenzi and Marcello Bertoldi, Analytical Methods for Markov Semigroups (2006)

M. A. Al-Gwaiz and S. A. Elsanousi, Elements of Real Analysis (2006)

Theodore G. Faticoni, Direct Sum Decompositions of Torsion-Free Finite Rank Groups (2007)

R. Sivaramakrishnan, Certain Number-Theoretic Episodes in Algebra (2006)

Aderemi Kuku, Representation Theory and Higher Algebraic K-Theory (2006)

Robert Piziak and P. L. Odell, Matrix Theory: From Generalized Inverses to Jordan Form (2007)

Norman L. Johnson, Vikram Jha, and Mauro Biliotti, Handbook of Finite Translation Planes (2007)

Lieven Le Bruyn, Noncommutative Geometry and Cayley-smooth Orders (2008)

Fritz Schwarz, Algorithmic Lie Theory for Solving Ordinary Differential Equations (2008)

Jane Cronin, Ordinary Differential Equations: Introduction and Qualitative Theory, Third Edition (2008)

Su Gao, Invariant Descriptive Set Theory (2009)

Christopher Apelian and Steve Surace, Real and Complex Analysis (2010)

Norman L. Johnson, Combinatorics of Spreads and Parallelisms (2010)

Lawrence Narici and Edward Beckenstein, Topological Vector Spaces, Second Edition (2010)

Moshe Sniedovich, Dynamic Programming: Foundations and Principles, Second Edition (2010)

Drumi D. Bainov and Snezhana G. Hristova, Differential Equations with Maxima (2011)

Willi Freeden, Metaharmonic Lattice Point Theory (2011)

Murray R. Bremner, Lattice Basis Reduction: An Introduction to the LLL Algorithm and Its Applications (2011)

Clifford Bergman, Universal Algebra: Fundamentals and Selected Topics (2011)

A. A. Martynyuk and Yu. A. Martynyuk-Chernienko, Uncertain Dynamical Systems: Stability and Motion Control (2012)

Washek F. Pfeffer, The Divergence Theorem and Sets of Finite Perimeter (2012)

Willi Freeden and Christian Gerhards, Geomathematically Oriented Potential Theory (2013)

Anatoly Martynyuk, Larisa Chernetskaya, and Vladislav Martynyuk, Weakly Connected Nonlinear Systems: Boundedness and Stability of Motion (2013)

Franz Halter-Koch, Quadratic Irrationals: An Introduction to Classical Number Theory (2013)

Łukasz Piasecki, Classification of Lipschitz Mappings (2014)

Classification of
Lipschitz Mappings

Łukasz Piasecki

CRC Press
Taylor & Francis Group
Boca Raton London New York

CRC Press is an imprint of the
Taylor & Francis Group, an **informa** business
A CHAPMAN & HALL BOOK

CRC Press
Taylor & Francis Group
6000 Broken Sound Parkway NW, Suite 300
Boca Raton, FL 33487-2742

First issued in paperback 2019

ISBN-13: 978-1-4665-9521-7 (hbk)
ISBN-13: 978-0-367-37904-9 (pbk)

Visit the Taylor & Francis Web site at
http://www.taylorandfrancis.com

and the CRC Press Web site at
http://www.crcpress.com

To My Parents

Contents

Introduction

The Lipschitz condition is of great importance in many branches of mathematics. The standard situation is the following: let (M, ρ) be a metric space; we say that a mapping $T : M \to M$ is lipschitzian if there exists a constant k such that, for all $x, y \in M$, we have

$$\rho(Tx, Ty) \le k\rho(x, y).$$

The class of all mappings satisfying the Lipschitz condition with a constant k is denoted by $L(k)$. The smallest constant k for which the above inequality holds is called the Lipschitz constant for T and is denoted by $k(T)$.

In some approaches, we also investigate a behavior of Lipschitz constants $k(T^n)$ for iterates T^n of T. To describe such behavior, we can find in the literature two special constants:

$$k_\infty(T) = \limsup_{n \to \infty} k(T^n)$$

and

$$k_0(T) = \lim_{n \to \infty} \sqrt[n]{k(T^n)}.$$

The constant $k_\infty(T)$ plays a special role in the fixed point theory for uniformly lipschitzian mappings; we say that a mapping $T : M \to M$ is uniformly lipschitzian if there exists a constant $k \ge 0$ such that, for every $n = 1, 2, \ldots$ and all $x, y \in M$, $\rho(T^n x, T^n y) \le k\rho(x, y)$.

The constant $k_0(T)$, which in case of linear mapping T defined on a Banach space is just the spectral radius of T, has the following interpretation in a general case of lipschitzian mappings:

$$k_0(T) = \inf k_r(T),$$

where $k_r(T)$ denotes the Lipschitz constant for T with respect to the metric r, and the infimum is taken over all metrics r that are equivalent to ρ; we say that a metric r is equivalent to ρ if there exist constants $a, b > 0$ such that, for all $x, y \in M$, $a\rho(x, y) \le r(x, y) \le b\rho(x, y)$.

In general, it is hard to determine or give nice evaluations for the constants $k_0(T)$ and $k_\infty(T)$ because it demands nice estimates of $k(T^n)$. The possible growth of the sequence $k(T^n)$ can be regulated by the following standard inequality:

$$k(T^n) \le k(T)^n \text{ for } n = 1, 2, \ldots,$$

1

which, in case of any class $L(k)$, is sharp. However, in many cases, this estimate doesn't give good approximation of $k(T^n)$. The sequence $\{k(T^n)\}$ may behave very unpredictably even in simple situations of functions defined on the interval $[0, 1]$, as seen below.

Let us consider two "similar" graphs. The first graph describes the function f:

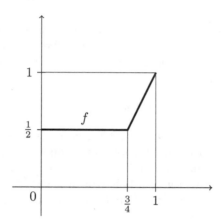

It is easy to check that $k(f) = 2$ and $k(f^n) = 2^n$ for $n = 1, 2, \ldots$. However, if we slightly change the graph of function f, then the situation will totally change:

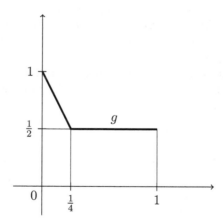

In this case, we can divide the interval $[0, 1]$ into two regions: $[0, 1/4]$ and $(1/4, 1]$. It is easy to verify that the function g extends a distance between any two points in $[0, 1/4]$, $|g(s) - g(t)| = 2\,|s - t|$, sending them to the interval

$(1/4, 1]$ on which the second iterate contracts it back, sending them to $\{\frac{1}{2}\}$. Consequently, $k(g) = 2$, but $k(g^n) = 0$ for $n \geq 2$.

In the above samples, we have $k(f) = k(g) = 2$. Hence, $f, g \in L(2)$. Nevertheless, remaining iterates behave in various ways: $k_\infty(f) = \infty$, $k_0(f) = 2$, and $k_\infty(g) = 0$, $k_0(g) = 0$.

In this book we deal with a problem of more precise classification of lipschitzian mappings. It seems natural that a condition that describes such classes should satisfy several principles. It should regulate the possible growth of the sequence of Lipschitz constants $k(T^n)$, as well as ensure nice estimates for $k_0(T)$ and $k_\infty(T)$. We also expect that such condition will provide some new results in the metric fixed point theory. The last and the most important request is that it has to be relatively easy to check.

In this monograph, we widely study the *mean Lipschitz condition*, which was introduced by Goebel, Japón Pineda, and Sims: suppose that $\alpha = (\alpha_1, \ldots, \alpha_n)$, where $n \geq 1$, $\alpha_i \geq 0$, $\alpha_1 > 0$, $\alpha_n > 0$, and $\sum_{i=1}^n \alpha_i = 1$; we say that $T : M \to M$ is α-lipschitzian for the constant $k \geq 0$ if, for all $x, y \in M$,

$$\sum_{i=1}^n \alpha_i \rho(T^i x, T^i y) \leq k\rho(x, y).$$

The class of all mappings that satisfies the above inequality with $\alpha = (\alpha_1, \ldots, \alpha_n)$ and a constant $k \geq 0$ is denoted by $L(\alpha, k)$.

We also focus on mean Lipschitz condition described by averages of order $p \geq 1$: let $\alpha = (\alpha_1, \ldots, \alpha_n)$ be as above and $p \geq 1$; we say that a mapping $T : M \to M$ is (α, p)-lipschitzian for the constant $k \geq 0$ if, for all $x, y \in M$,

$$\left(\sum_{i=1}^n \alpha_i \rho(T^i x, T^i y)^p\right)^{1/p} \leq k\rho(x, y).$$

The class of all mappings that satisfies the above condition with $\alpha = (\alpha_1, \ldots, \alpha_n)$, $p \geq 1$ and $k \geq 0$ is denoted by $L(\alpha, p, k)$. Let us note that, for $p = 1$, we get a definition of α-lipschitzian mapping for the constant k. We usually write $L(\alpha, k)$ instead of $L(\alpha, 1, k)$. It is clear that, for $n = 1$, we get nothing more than the classical definition of k-lipschitzian mapping. Hence, in this special case, the mean Lipschitz condition determines the class $L(k)$.

The presented monograph arose as a result of discussions that took place during a seminar under the supervision of Professor Kazimierz Goebel at the University of Maria Curie-Sklodowska in Lublin, Poland. The included results have been presented many times during international conferences as well as at seminars devoted to the metric fixed point theory. Many of them have been already published. A part of the presented material appeared only in a PhD thesis of the present author [70] written under the guidance of Professor Kazimierz Goebel.

The book is addressed to advanced undergraduate and graduate students as well as to professionals looking for new topics in the metric fixed point theory. Nevertheless, we deeply believe that this monograph will be of interest among readers working on other areas (for example, differential equations and dynamical systems). The proposed text is self-contained, and only a basic knowledge of functional analysis and topology is required.

The monograph covers approximately 220 pages of a systematic course, with updated facts concerning discussed topics, with accompanying illustrations, a rich collection of examples, and open problems. The book does not include typical exercises. Instead, we can find many sentences like "It is easy to see...", "We leave to the reader...", "Observe that.." and so forth.

This book does not aim to compete with any book devoted to the metric fixed point theory. Our goal is to provide a systematic, self-contained course of a new classification of Lipschitz mappings, pointing out its application in many topics of the metric fixed point theory. Moreover, we encourage new adept to acquaint themselves with some existing books, in particular [49], [36] and [31], which, with no doubt, could be useful for further development of this new field.

I am deeply grateful to Professor Kazimierz Goebel for creating the atmosphere conducive to writing this book. His constant interest in progress, support, and discussions while writing were valuable to me and undoubtedly enriched the content of this monograph considerably.

It is my kind duty to acknowledge Professors Tomás Domínguez Benavides, Emanuele Casini, and Stanisław Prus for pointing out errors and omissions as well as for valuable remarks that improved this book.

I would like to express my deep sense of gratitude to my colleague Dr. Piotr Oleszczuk for his kind support in proofreading and verifying the substantial content of subsequent parts of the text.

I would like to express special thanks to my friend and scientific partner, Dr. Víctor Pérez García. During his postdoctoral position at Maria Curie-Sklodowska University, we had a number of opportunities to hold discussions as well as intensive and fruitful cooperation. Many results obtained then were included in this monograph.

I am thankful to Sunil Nair for excellent cooperation on publishing this book.

Chapter 1

The Lipschitz condition

Let (M, ρ) be a metric space. By $B_M(x, r)$, we denote *the closed ball* centered at $x \in M$ with radius $r > 0$:

$$B_M(x, r) = \{y \in M : \rho(x, y) \leq r\}.$$

By $S_M(x, r)$, we denote *the sphere* centered at $x \in M$ with radius $r > 0$:

$$S_M(x, r) = \{y \in M : \rho(x, y) = r\}.$$

As usual, we shall drop the subscript when M is clear from the context. For a subset A of (M, ρ), $\text{diam}(A)$ denotes *the diameter* of A; that is,

$$\text{diam}(A) = \sup\{\rho(x, y) : x, y \in A\}.$$

A subset $A \subset M$ is called *bounded* if $\text{diam}(A) < \infty$.

We say that $z \in M$ is a *fixed point* of mapping $T : M \to M$ if $z = Tz$. The set of all fixed points of T is denoted by $\text{Fix}(T)$.

A mapping $T : M \to M$ is said to be *lipschitzian* if there exists $k \geq 0$ such that, for all $x, y \in M$

$$\rho(Tx, Ty) \leq k\rho(x, y). \tag{1.1}$$

If we want to indicate $k \geq 0$, then we refer to such a mapping as k-*lipschitzian*. By $L(k)$, we denote the class of all lipschitzian mappings that satisfies (1.1) with a constant k. It is clear that lipschitzian mappings are uniformly continuous. For a given mapping T, the smallest k for which (1.1) holds (such k always exists) is called the *Lipschitz constant* for T and will be denoted by $k_\rho(T)$ or simply $k(T)$ when the underlying metric is clear from the context. A mapping $T : M \to M$ is said to be a *contraction* if $k(T) < 1$. If $k(T) \leq 1$, then T is called *nonexpansive*. Thus, the class of nonexpansive mappings includes all contractions and isometries, in particular, the identity.

1.1 Nonlinear spectral radius

For two lipschitzian mappings $T, S : M \to M$ and all $x, y \in M$, we have

$$\rho(TSx, TSy) \leq k(T)\rho(Sx, Sy) \leq k(T)k(S)\rho(x, y)$$

and so
$$k(T \circ S) \leq k(T)k(S). \tag{1.2}$$

In particular, (1.2) regulates the possible growth of the sequence of Lipschitz constants for iterates T^n of T,

$$k(T^{n+m}) = k(T^n T^m) \leq k(T^n) \cdot k(T^m) \text{ for } m, n = 1, 2, \ldots$$

and consequently
$$k(T^n) \leq k(T)^n \text{ for } n = 1, 2, \ldots. \tag{1.3}$$

Thus, if T is nonexpansive, then all its powers T^n are also nonexpansive. If T is a contraction, then $\lim_{n \to \infty} k(T^n) = 0$. To recall another basic observation about the sequence of Lipschitz constants $\{k(T^n)\}$ of a given mapping T, we shall need the following well-known fact:

Lemma 1.1 *Let $\{a_n\}_{n=1}^{\infty}$ be a sequence of nonnegative numbers satisfying, for all $m, n = 1, 2, \ldots$, the following condition:*

$$a_{m+n} \leq a_m \cdot a_n. \tag{1.4}$$

Then, $\lim\limits_{n \to \infty} \sqrt[n]{a_n} = \inf \left\{ \sqrt[n]{a_n} : n = 1, 2, \ldots \right\}$.

Proof. Put
$$a = \inf \left\{ \sqrt[n]{a_n} : n = 1, 2, \ldots \right\}.$$

Fix $\epsilon > 0$. There is $k \in \mathbb{N}$ such that $a \leq \sqrt[k]{a_k} \leq a + \epsilon$. For any positive integer n, we can write
$$n = qk + r,$$

where q and r are nonnegative integers with $0 \leq r \leq k - 1$. Then, using (1.4), we get

$$a_n \leq a_{qk} \cdot a_r \leq a_k^q \cdot a_r.$$

Thus,
$$\sqrt[n]{a_n} \leq a_k^{q/n} \cdot a_r^{1/n} \leq (a + \epsilon)^{kq/n} \cdot a_r^{1/n}.$$

Letting n tend to infinity and for fixed k we obtain

$$\lim_{n \to \infty} \frac{kq}{n} = \lim_{n \to \infty} \left(1 - \frac{r}{n}\right) = 1,$$

and consequently,
$$a \leq \limsup_{n \to \infty} \sqrt[n]{a_n} \leq a + \epsilon.$$

Since $\epsilon > 0$ is arbitrary, this implies $\limsup\limits_{n \to \infty} \sqrt[n]{a_n} = a$. On the other hand, $\sqrt[n]{a_n} \geq a$ for each n and so $\liminf\limits_{n \to \infty} \sqrt[n]{a_n} \geq a$. Thus, $\lim\limits_{n \to \infty} \sqrt[n]{a_n} = a$.

$\square$

We have already seen that the sequence of Lipschitz constants $\{k(T^n)\}$ satisfies (1.4). Thus, the constant

$$k_0(T) = \lim_{n\to\infty} \sqrt[n]{k(T^n)}$$

is well defined and

$$k_0(T) = \lim_{n\to\infty} \sqrt[n]{k(T^n)} = \inf\left\{ \sqrt[n]{k(T^n)} : n = 1, 2, \ldots \right\}. \qquad (1.5)$$

Using (1.3), we obtain the following evaluation for $k_0(T)$:

$$k_0(T) = \lim_{n\to\infty} \sqrt[n]{k(T^n)} \le k(T).$$

We say that two metrics d and ρ are *equivalent* if there exist two constants $a > 0$ and $b > 0$ such that, for all $x, y \in M$

$$ad(x, y) \le \rho(x, y) \le bd(x, y).$$

It is easy to check that any mapping T, which is lipschitzian with respect to a given metric ρ is also lipschitzian with respect to any equivalent metric d. The respective Lipschitz constants $k_\rho(T)$ and $k_d(T)$ may differ and the difference is regulated by the relation

$$d(Tx, Ty) \le \frac{1}{a}\rho(Tx, Ty) \le \frac{1}{a}k_\rho(T)\rho(x, y) \le \frac{b}{a}k_\rho(T)d(x, y).$$

This implies that

$$k_d(T) \le \frac{b}{a}k_\rho(T).$$

Changing roles of metrics d and ρ, we obtain

$$k_\rho(T) \le \frac{b}{a}k_d(T).$$

Obviously, the same holds for all iterates T^n of mapping T; hence,

$$\frac{a}{b}k_\rho(T^n) \le k_d(T^n) \le \frac{b}{a}k_\rho(T^n). \qquad (1.6)$$

In particular, (1.6) implies that the constant $k_0(T)$ is independent of the selection of a metric d, which is equivalent to ρ. Thus, using (1.5), for each equivalent metric d, we have $k_0(T) \le k_d(T)$. As a consequence, $k_0(T) \le \inf k_d(T)$, where infimum is taken over all metrics d, which are equivalent to ρ. Actually, we have $k_0(T) = \inf k_d(T)$. Indeed, for arbitrary $\epsilon > 0$ set $\lambda = 1/(k_0(T) + \epsilon)$. Let us define a metric d_λ by putting for each $x, y \in M$

$$d_\lambda(x, y) = \sum_{i=0}^{\infty} \rho(T^i x, T^i y)\lambda^i = \rho(x, y) + \sum_{i=1}^{\infty} \rho(T^i x, T^i y)\lambda^i. \qquad (1.7)$$

It is easy to observe that the metric d_λ is equivalent to ρ and

$$\rho(x,y) \le d_\lambda(x,y) \le \left(\sum_{i=0}^{\infty} k_\rho(T^i)\lambda^i \right) \rho(x,y). \tag{1.8}$$

Now, for all $x, y \in M$, we have

$$
\begin{aligned}
d_\lambda(Tx, Ty) &= \sum_{i=0}^{\infty} \rho(T^{i+1}x, T^{i+1}y)\lambda^i \\
&= \frac{1}{\lambda} \sum_{i=0}^{\infty} \rho(T^{i+1}x, T^{i+1}y)\lambda^{i+1} \\
&\le \frac{1}{\lambda} \sum_{i=0}^{\infty} \rho(T^ix, T^iy)\lambda^i \\
&= \frac{1}{\lambda} d_\lambda(x,y).
\end{aligned}
$$

Thus,

$$k_{d_\lambda}(T) \le 1/\lambda = k_0(T) + \epsilon. \tag{1.9}$$

Since $\epsilon > 0$ is arbitrarily chosen, we finally obtain the following characterization of $k_0(T)$:

$$k_0(T) = \inf \{ k_d(T) : \ d \text{ is equivalent to } \rho \}. \tag{1.10}$$

1.2   Uniformly lipschitzian mappings

Of particular interest in the metric fixed point theory is a class of *uniformly lipschitzian mappings*. We say that a mapping $T : M \to M$ is *uniformly lipschitzian*, if there exists a constant $k \ge 0$ such that for all $x, y \in M$ and $n = 1, 2, \ldots$, we have

$$\rho(T^n x, T^n y) \le k\rho(x,y).$$

If we want to indicate $k \ge 0$, then we refer to such mapping as a *uniformly k-lipschitzian*. Equivalently, a lipschitzian mapping T is uniformly lipschitzian, if and only if

$$\sup \{ k(T^n) : n = 1, 2, \ldots \} < \infty. \tag{1.11}$$

We shall denote the class of all uniformly k-lipschitzian mappings by $L_u(k)$.

It is well known that there exists another characterization of such mappings. It is clear that, if for some equivalent metric d we have $k_d(T) \le 1$, then $k_\rho(T^n) \le b/a$ for $n = 1, 2, \ldots$ (a and b are taken from the definition of equivalent metrics). Hence, condition (1.11) is satisfied. On the other hand, if T is uniformly lipschitzian, then it is nonexpansive with respect to the equivalent metric defined by

$$d(x, y) = \sup \{\rho(T^n x, T^n y) : n = 0, 1, 2, \ldots\}. \tag{1.12}$$

Indeed, for all $x, y \in M$, we have

$$\rho(x, y) \leq d(x, y) \leq \sup \{k(T^n) : n = 0, 1, 2, \ldots\} \rho(x, y)$$

and

$$
\begin{aligned}
d(Tx, Ty) &= \sup \{\rho(T^{n+1}x, T^{n+1}y) : n = 0, 1, 2, \ldots\} \\
&\leq \sup \{\rho(T^n x, T^n y) : n = 0, 1, 2, \ldots\} \\
&= d(x, y).
\end{aligned}
$$

Finally, we conclude that a mapping T is uniformly lipschitzian if and only if there exists an equivalent metric with respect to which T is nonexpansive. Consequently, for any uniformly lipschitzian mapping T, we have $k_0(T) \leq 1$.

Also, there is another useful constant that appears during studies devoted to uniformly lipschitzian mappings. For any lipschitzian mapping $T : M \to M$, we define the constant $k_\infty(T)$ as

$$k_\infty(T) = \limsup_{n \to \infty} k(T^n).$$

Obviously, a mapping T is uniformly lipschitzian if and only if $k_\infty(T) < \infty$. Usefulness of the constant $k_\infty(T)$ follows from the fact that usually

$$k_\infty(T) < \sup \{k(T^n) : n = 1, 2, \ldots\}.$$

We only mention that all the above definitions carry over to a Banach space setting $(X, \|\cdot\|)$ by taking $\rho(x, y) = \|x - y\|$. It is easy to verify that, for any nonempty subset C of a Banach space X and two lipschitzian mappings $T, S : C \to X$, we additionally have the following elementary inequalities: $k(T + S) \leq k(T) + k(S)$ and $k(\lambda T) = \lambda k(T)$ for $\lambda \geq 0$.

In this text, unless otherwise stated, we always assume that all considered sets are nonempty.

Chapter 2

Basic facts on Banach spaces

Let $(X, \|\cdot\|)$ be a real Banach space. In this case, the *closed unit ball*

$$B_X(0,1) = \{x \in X : \|x\| \leq 1\}$$

is denoted by B_X and the *unit sphere*

$$S_X(0,1) = \{x \in X : \|x\| = 1\}$$

by S_X. If X is clear from this context, then we write B and S instead of B_X and S_X, respectively.

Let us introduce some notations of classical Banach spaces.

- For each $1 \leq p < \infty$, ℓ_p denotes the space that consists of all p-summable sequences $x = (x_1, x_2, \dots)$ of real numbers furnished with the standard norm

$$\|x\| = \left(\sum_{i=1}^{\infty} |x_i|^p \right)^{\frac{1}{p}}.$$

- ℓ_∞ is the Banach space that consists of all bounded sequences $x = (x_1, x_2, \dots)$ of real numbers, c denotes its subspace of all converging sequences, and c_0 the subspace of c of all null sequences. All of them are furnished with the standard sup norm,

$$\|x\| = \sup\{|x_i| : i = 1, 2, \dots\}.$$

  In the case of c_0 space, the above formula can be written as

$$\|x\| = \max\{|x_i| : i = 1, 2, \dots\}.$$

- $C[0,1]$ denotes the space that consists of all continuous functions $f : [0,1] \to \mathbb{R}$ with the sup norm,

$$\|f\| = \sup\{|f(t)| : t \in [0,1]\} = \max\{|f(t)| : t \in [0,1]\}.$$

- For each $1 \leq p < \infty$ by $L_p[0,1]$, we denote the space of all Lesbesgue measurable functions $f : [0,1] \to \mathbb{R}$ such that $\int_{[0,1]} |f(t)|^p \, dt < \infty$, identifying those functions that differ only on a null set, and the norm is defined as

$$\|f\| = \left(\int_{[0,1]} |f(t)|^p \, dt \right)^{\frac{1}{p}}.$$

- $L_\infty[0,1]$ is the space of all Lesbesgue measurable functions $f : [0,1] \to \mathbb{R}$ for which

$$\|f\| = \sup \text{ ess} \{|f(t)| : t \in [0,1]\} = \inf_{A \in \mathcal{N}} \left(\sup_{t \in [0,1]\backslash A} |f(t)| \right),$$

where $\mathcal{N} = \{A \subset [0,1] : A \text{ is a null set}\}$. Here, as in the previous case of $L_p[0,1]$ spaces, we identify functions that differ only on a set of measure zero.

2.1   Convexity

Let X be a Banach space with the closed unit ball B_X and the unit sphere S_X. A subset K of X is said to be *convex* if, for any $x,y \in K$ and any $\lambda \in [0,1]$, the point $(1-\lambda)x + \lambda y$ belongs to K. The unit ball B_X and, consequently, any ball in X is convex. Indeed, if $x,y \in B_X$, then for any $\lambda \in [0,1]$,

$$\|(1-\lambda)x + \lambda y\| \leq (1-\lambda)\|x\| + \lambda\|y\| \leq 1.$$

It is easy to see that, for any $z \in X$ and for any $r > 0$, we have

$$B_X(z,r) = z + rB_X,$$

where $rB_X = \{rx : x \in B_X\}$.

For any nonempty set $A \subset X$ by $\text{conv}(A)$, we denote the *convex hull* of A, that is, the smallest convex set containing A. Hence,

$$\text{conv}(A) = \bigcap \{K \subset X : K \supset A, K \text{ is convex set}\}.$$

It means that $x \in \text{conv}(A)$ if and only if $x = \sum_{i=1}^n \lambda_i x_i$ for some $x_1, \ldots, x_n \in A$ and some scalars $\lambda_i \geq 0$ for which $\sum_{i=1}^n \lambda_i = 1$. Obviously, A is convex if and only if $A = \text{conv}(A)$. If A is convex, then its closure $\overline{A}$ is also convex. Clearly, $B_X = \text{conv}(S_X)$.

The set

$$\overline{\text{conv}}(A) = \bigcap \{K \subset X : K \supset A, K \text{ is convex and closed in } X\}$$

is said to be the *convex closure* of A. Equivalently, $\overline{\text{conv}}(A)$ is the smallest convex and closed set containing A. In other words, $\overline{\text{conv}}(A) = \overline{\text{conv}(A)}$.

It is clear that

$$A \subset \text{conv}(A) \subset \overline{\text{conv}}(A),$$

and it is easy to prove that, for any bounded set A in X,

$$\text{diam}(A) = \text{diam}(\text{conv}(A)) = \text{diam}(\overline{\text{conv}}(A)).$$

The following very important result states that compactness is invariant under taking a convex closure.

Theorem 2.1 (Mazur) *If $\overline{A}$ is compact, then $\overline{conv}(A)$ is also compact.*

We say that a Banach space X is *strictly convex* if, for all $x, y \in X$, the following implication holds:

$$\left.\begin{array}{c} \|x\| \leq 1 \\ \|y\| \leq 1 \\ \|x - y\| > 0 \end{array}\right\} \Rightarrow \left\|\frac{x + y}{2}\right\| < 1.$$

The above condition means that a sphere does not contain a segment that is not a singleton. In such space, any three points x, z, y such that $\|x - z\| + \|z - y\| = \|x - y\|$ must lie on a line. More precisely,

$$z = \frac{\|y - z\|}{\|x - y\|} x + \frac{\|x - z\|}{\|x - y\|} y.$$

Equivalently, X is strictly convex if and only if the following implication holds: for each $x, y \in X$

$$(x \neq 0 \wedge y \neq 0 \wedge y \neq tx \text{ for each } t > 0) \Longrightarrow (\|x + y\| < \|x\| + \|y\|).$$

We say that a Banach space $(X, \|\cdot\|)$ is *uniformly convex* if, for each $\epsilon \in (0, 2]$, there exists $\delta > 0$ such that the following implication holds: for each $x, y \in X$

$$\left.\begin{array}{c} \|x\| \leq 1 \\ \|y\| \leq 1 \\ \|x - y\| \geq \epsilon \end{array}\right\} \Rightarrow \left\|\frac{x + y}{2}\right\| \leq 1 - \delta. \tag{2.1}$$

One can check that X is uniformly convex if and only if for any two sequences $\{x_n\}$ and $\{y_n\}$ in X the following implication holds:

$$\left(\lim_{n \to \infty} \|x_n\| = \lim_{n \to \infty} \|y_n\| = 1 \wedge \lim_{n \to \infty} \|x_n + y_n\| = 2\right) \Rightarrow \lim_{n \to \infty} \|x_n - y_n\| = 0.$$

The condition of uniformly convexity was introduced by Clarkson [20] in 1936, and his classical result states that all Hilbert spaces, all $L_p[0, 1]$, and ℓ_p spaces with $1 < p < \infty$ are uniformly convex.

The *modulus of convexity* of a Banach space X is the function $\delta_X : [0, 2] \to [0, 1]$ defined by

$$\delta_X(\epsilon) = \inf\left\{1 - \frac{\|x + y\|}{2} : \|x\| \leq 1, \|y\| \leq 1, \|x - y\| \geq \epsilon\right\}.$$

Hence, for any $\epsilon > 0$, the number $\delta_X(\epsilon)$ is the largest number for which the implication (2.1) holds. It is easy to note that X is uniformly convex if and only if $\delta_X(\epsilon) > 0$ for every $\epsilon > 0$. Furthermore, for all $x, y \in X$ and any $r > 0$, the following implication is satisfied:

$$\left.\begin{array}{l} \|x\| \leq r \\ \|y\| \leq r \end{array}\right\} \Rightarrow \left\|\frac{x+y}{2}\right\| \leq \left(1 - \delta_X\left(\frac{\|x-y\|}{r}\right)\right) r. \qquad (2.2)$$

The *characteristic of convexity* of a Banach space X is the number

$$\epsilon_0(X) = \sup\left\{\epsilon \in [0,2] : \delta_X(\epsilon) = 0\right\}.$$

Hence, the number $\epsilon_0(X)$ bounds the length of segments lying on the unit sphere S or arbitrarily close to it. Obviously, $\epsilon_0(X) = 0$ if and only if X is uniformly convex.

Below we list several very important properties of the modulus of convexity δ_X and the characteristic of convexity $\epsilon_0(X)$. The proofs can be found in [36].

- $\delta_X(0) = 0$ and δ_X is nondecreasing on $[0,2]$.

- $\delta_X(2) = 1$ if and only if X is strictly convex.

- δ_X is continuous on $[0,2)$.

However, it may happen that δ_X is not continuous at $\epsilon = 2$. Such a situation takes place in example 2.2. In spite of this, we always have

- $\lim_{\epsilon \to 2^-} \delta(\epsilon) = 1 - \frac{\epsilon_0(X)}{2}$.

In addition,

- δ_X is strictly increasing on $[\epsilon_0(X), 2]$.

It is easy to check that a Hilbert space H is uniformly convex as a consequence of the **Parallelogram Law**

$$\|x+y\|^2 + \|x-y\|^2 = 2\left(\|x\|^2 + \|y\|^2\right),$$

which also implies that

$$\delta_H(\epsilon) = 1 - \sqrt{1 - \frac{\epsilon^2}{4}}.$$

In the case of ℓ_p and L_p, $2 < p < \infty$, we have

$$\delta_p(\epsilon) = 1 - \left(1 - \left(\frac{\epsilon}{2}\right)^p\right)^{\frac{1}{p}}.$$

If $1 < p < 2$, then we have the following implicit formula obtained by Hanner in 1956:

$$\left|1 - \delta_p(\epsilon) + \frac{\epsilon}{2}\right|^p + \left|1 - \delta_p(\epsilon) - \frac{\epsilon}{2}\right|^p = 2.$$

Example 2.1 *Consider two-dimensional space $\ell_1^{(2)}$ furnished with the standard norm,*

$$\|(x_1, x_2)\| = |x_1| + |x_2|.$$

If $e_1 = (1,0)$ and $e_2 = (0,1)$, then the whole segment

$$[e_1, e_2] = \{(1 - \lambda)e_1 + \lambda e_2 : \lambda \in [0,1]\}$$

lies on the unit sphere S. Consequently,

$$\delta_{\ell_1^{(2)}}(\epsilon) = 0 \text{ for any } \epsilon \in [0,2].$$

Furthermore, since the spaces ℓ_1, ℓ_∞, c_0, c, $C[0,1]$, $L_1[0,1]$, $L_\infty[0,1]$ contain an isometric copy of $\ell_1^{(2)}$, we conclude that $\delta_X(\epsilon) = 0$ for any $\epsilon \in [0,2]$ if X is one of them.

It is clear that each uniformly convex space is also strictly convex. In finite dimensional Banach spaces, both conditions are equivalent. It is an easy consequence of the fact that a closed ball is compact. In infinite dimensional spaces, such equivalence does not hold.

Example 2.2 *Consider the space c_0 furnished with the standard norm given for any $x = (x_1, x_2, \ldots)$ by*

$$\|x\| = \sup\{|x_i| : i = 1, 2, \ldots\}.$$

Define the new norm

$$\||x\|| = \left(\|x\|^2 + \sum_{i=1}^{\infty}\left(\frac{x_i}{i}\right)^2\right)^{\frac{1}{2}}.$$

It is easy to observe that both norms are equivalent and

$$\|x\| \leq \||x\|| \leq \left(1 + \frac{\pi^2}{6}\right)^{\frac{1}{2}}\|x\|.$$

The space $(c_0, \||\cdot\||)$ is strictly convex. To see it, consider any two points $x = (x_1, x_2, \ldots)$ and $y = (y_1, y_2, \ldots)$ in c_0 such that $x \neq 0$, $y \neq 0$, and $y \neq tx$ for each $t > 0$. If we now denote

$$u = \left(\|x\|, \frac{x_1}{1}, \frac{x_2}{2}, \ldots\right),$$

$$v = \left(\|y\|, \frac{y_1}{1}, \frac{y_2}{2}, \ldots\right),$$

then also $u \neq 0$, $v \neq 0$, and $v \neq tu$ for each $t > 0$. Consequently, by the triangular inequality, both for the supremum norm and for the Euclidean norm, we obtain

$$\||x+y\|| = \left(\|x+y\|^2 + \sum_{i=1}^{\infty}\left(\frac{x_i+y_i}{i}\right)^2\right)^{\frac{1}{2}}$$

$$\leq \left((\|x\|+\|y\|)^2 + \sum_{i=1}^{\infty}\left(\frac{x_i}{i}+\frac{y_i}{i}\right)^2\right)^{\frac{1}{2}}$$

$$< \left(\|x\|^2 + \sum_{i=1}^{\infty}\left(\frac{x_i}{i}\right)^2\right)^{\frac{1}{2}} + \left(\|y\|^2 + \sum_{i=1}^{\infty}\left(\frac{y_i}{i}\right)^2\right)^{\frac{1}{2}}$$

$$= \||x\|| + \||y\||,$$

as we desired.

Let

$$u_n = (\underbrace{0,\ldots,0}_{n-1},1,1,0,\ldots,0,\ldots)$$

and

$$v_n = (\underbrace{0,\ldots,0}_{n-1},-1,1,0,\ldots,0,\ldots).$$

Then,

$$\||u_n\|| = \||v_n\|| = \left(1+\frac{1}{n^2}+\frac{1}{(n+1)^2}\right)^{\frac{1}{2}},$$

$$\||u_n-v_n\|| = \||(\underbrace{0,\ldots,0}_{n-1},2,0,\ldots,0,\ldots)\|| = \left(4+\left(\frac{2}{n}\right)^2\right)^{\frac{1}{2}}$$

and

$$\||u_n+v_n\|| = \||(\underbrace{0,\ldots,0}_{n},2,0,\ldots,0,\ldots)\|| = \left(4+\frac{4}{(n+1)^2}\right)^{\frac{1}{2}}.$$

We have

$$\lim_{n\to\infty}\||u_n\|| = \lim_{n\to\infty}\||v_n\|| = 1, \quad \lim_{n\to\infty}\||u_n+v_n\|| = 2$$

and

$$\lim_{n\to\infty}\||u_n-v_n\|| = 2;$$

hence, $(c_0,\||\cdot\||)$ is not uniformly convex. Moreover,

$$\delta_{(c_0,\||\cdot\||)}(\epsilon) = \begin{cases} 0 & \text{if } \epsilon \in [0,2), \\ 1 & \text{if } \epsilon = 2. \end{cases}$$

2.2 The operator norm

For any two Banach spaces $(X, \|\cdot\|)$ and $(Y, \|\cdot\|)$ by $\mathfrak{L}(X, Y)$, we denote the space of all norm-to-norm continuous linear operators $T : X \to Y$. If $X = Y$, then we write $\mathfrak{L}(X)$ instead of $\mathfrak{L}(X, X)$. For any $T \in \mathfrak{L}(X, Y)$, we define the *norm* $\|\cdot\|$ *of* T as

$$\|T\| = \sup \left\{ \|Tx\| : x \in B_X \right\}.$$

It is well known that the norm of T can be equivalently given by

$$
\begin{aligned}
\|T\| &= \sup \left\{ \|Tx\| : x \in S_X \right\} \\
&= \sup \left\{ \frac{\|Tx\|}{\|x\|} : x \in X,\, x \neq 0 \right\} \\
&= \inf \left\{ M > 0 : \|Tx\| \leq M \|x\| \text{ for all } x \in X \right\}.
\end{aligned}
$$

One can observe that the last equation means that actually the norm of the operator T is just its Lipschitz constant. Hence, using our notations, we can write

$$\|T\| = k(T).$$

If $T \in \mathfrak{L}(X)$, then the estimate (1.3) can be written as

$$\|T^n\| \leq \|T\|^n,$$

and the relation (1.5), which now takes the form

$$k_0(T) = \lim_{n \to \infty} \sqrt[n]{\|T^n\|} = \inf \left\{ \sqrt[n]{\|T^n\|} : n = 1, 2, \dots \right\},$$

is just the *spectral radius* of T, which we denote by $\mathrm{spr}(T)$. Consequently,

$$k_0(T) = \mathrm{spr}(T) = \lim_{n \to \infty} \sqrt[n]{\|T^n\|} \leq \|T\|.$$

It is easy to verify that, if X is a normed space and Y is a Banach space, then $(\mathfrak{L}(X, Y), \|\cdot\|)$ is a Banach space. In the latter, we consider the special case when $(Y, \|\cdot\|) = (\mathbb{R}, |\cdot|)$, where $|\cdot|$ denotes the absolute value.

2.3 Dual spaces, reflexivity, the weak, and weak* topologies

Let $(X, \|\cdot\|)$ be a real Banach space with the closed unit ball B_X and the unit sphere S_X. Recall that a *dual space* or *conjugate space* X^* of X is the

Banach space that consists of all continuous linear functionals $x^* : X \to \mathbb{R}$ furnished with a *dual norm* $\|\cdot\|_*$ given by

$$\|x^*\|_* = \sup\left\{x^*(x) : x \in B_X\right\} = \sup\left\{x^*(x) : x \in S_X\right\}.$$

The *second conjugate* space X^{**} of X is defined as $X^{**} = (X^*)^*$.

The *canonical embedding* of X into X^{**} is the mapping $j : X \to X^{**}$ defined for any $x \in X$ and $x^* \in X^*$ by

$$j(x)(x^*) = x^*(x).$$

We leave for the reader to check that j is a linear mapping and $j(x) \in X^{**}$ for any $x \in X$. Moreover, j is an isometry; that is, for any $x \in X$, $\|j(x)\|_{**} = \|x\|$. This is one of the many useful consequences of the fundamental Hahn-Banach Theorem.

Theorem 2.2 (Hahn-Banach) *Let Y be a linear subspace of a Banach space X. Then for every continuous linear functional y^* on Y there exists a continuous linear functional $x^* \in X^*$ such that $x^*(y) = y^*(y)$ for every $y \in Y$ and*

$$\|x^*\|_* = \|y^*\|_* := \sup\left\{y^*(y) : y \in B_Y\right\}.$$

To prove that the canonical embedding j is an isometry, it is enough to observe that, in view of the Hahn-Banach Theorem, for any $x \in X$, $x \neq 0$, there exists $x^* \in X^*$ such that $x^*(x) = \|x\|$ and $\|x^*\|_* = 1$.

If $j(X) = X^{**}$, then X is said to be *reflexive*, and we usually write $X = X^{**}$. If j is not surjective, then X is called *nonreflexive*. If $\|\cdot\|_1$ and $\|\cdot\|_2$ are two *equivalent norms* on X, then $(X, \|\cdot\|_1)$ is reflexive if and only if $(X, \|\cdot\|_2)$ is reflexive; recall that two norms $\|\cdot\|_1$ and $\|\cdot\|_2$ on X are said to be equivalent if there exist constants $a, b > 0$ such that for each $x \in X$

$$a\|x\|_2 \leq \|x\|_1 \leq b\|x\|_2.$$

Each $x^* \in X^*$ generates a seminorm p_{x^*} on X defined for any $x \in X$ by

$$p_{x^*}(x) = |x^*(x)|.$$

The *weak topology* $\mathcal{T}(X, X^*)$ on X is the topology generated by the family of seminorms $\{p_{x^*} : x^* \in X^*\}$ described above. It means that $\mathcal{T}(X, X^*)$ is the weakest topology on X with respect to which all functionals $x^* \in X^*$ are continuous. Hence, if $\mathcal{T}(X)$ denotes the topology generated by the norm $\|\cdot\|$ on X, we have $\mathcal{T}(X, X^*) \subset \mathcal{T}(X)$. The weak topology coincides with the norm topology on X if and only if X is a finite dimensional Banach space. Furthermore, if X is infinite dimensional Banach space, then every nonempty weakly open set is unbounded; by a *weakly open* set, we understand any element of the weak topology $\mathcal{T}(X, X^*)$. We say that a subset K of X is *weakly closed* if its complement in X is weakly open set.

Similarly, each $x \in X$ generates a seminorm p_x on X^* defined for any $x^* \in X^*$ as

$$p_x(x^*) = |x^*(x)| = |j(x)(x^*)|.$$

The *weak* topology* $\mathcal{T}(X^*, X)$ on X^* is the topology generated by the family of seminorms $\{p_x : x \in X\}$. In other words, $\mathcal{T}(X^*, X)$ is the weakest topology on X^* with respect to which all functionals $x^{**} \in j(X)$ are continuous. Hence,

$$\mathcal{T}(X^*, X) \subset \mathcal{T}(X^*, X^{**}) \subset \mathcal{T}(X^*),$$

where $\mathcal{T}(X^*, X^{**})$ denotes the weak topology on X^* and $\mathcal{T}(X^*)$ denotes the topology generated by a norm $\|\cdot\|_*$ in X^*. Moreover,

$$\mathcal{T}(X^*, X) = \mathcal{T}(X^*, X^{**})$$

if and only if X is reflexive. We say that $K \subset X^*$ is *weakly* closed* if its complement in X^* belongs to $\mathcal{T}(X^*, X)$.

A sequence $\{x_n\} \subset X$ converges to a point $x \in X$ in the weak topology if and only if for every $x^* \in X^*$

$$\lim_{n \to \infty} x^*(x_n) = x^*(x).$$

Then we say that $\{x_n\}$ is *weakly convergent* or *converges weakly* to x, and we write

$$w - \lim_{n \to \infty} x_n = x.$$

Similarly, a sequence $\{x_n^*\} \subset X^*$ converges to a point $x^* \in X^*$ in the weak* topology if and only if for every $x \in X$

$$\lim_{n \to \infty} x_n^*(x) = x^*(x);$$

we say that $\{x_n^*\}$ is *weakly* convergent* to x^*, and we write

$$w^* - \lim_{n \to \infty} x_n^* = x^*.$$

A set $A \subset X$ is said to be *linearly dense* in X, if the set

$$\mathrm{span}(A) := \left\{ \sum_{i=1}^{n} \lambda_i x_i : x_1, \ldots, x_n \in A, \lambda_1, \ldots, \lambda_n \in \mathbb{R}, n \in \mathbb{N} \right\}$$

is *dense* in X; that is, $\overline{\mathrm{span}(A)} = X$.

We shall list some well-known facts concerning weak and weak* topologies.

- Let X be a Banach space and $A \subset X^*$ be a linearly dense set in X^*. A bounded sequence $\{x_n\} \subset X$ converges weakly to x in X if and only if $x^*(x) = \lim_{n \to \infty} x^*(x_n)$ for each $x^* \in A$.

- Let $A \subset X$ be a linearly dense in X. A bounded sequence $\{x_n^*\} \subset X^*$ converges weakly* to $x^* \in X^*$ if and only if $x^*(x) = \lim_{n \to \infty} x_n^*(x)$ for each $x \in A$.

- A convex set $K \subset X$ is weakly closed if and only if it is closed.

The last property follows immediately from the **Separation Theorem**, which is actually a consequence of the Hahn-Banach Theorem.

Theorem 2.3 *If K and C are disjoint convex subsets of X such that K is closed and C is compact, then there exists $x^* \in X^*$ such that*

$$\max \left\{ x^*(x) : x \in C \right\} < \inf \left\{ x^*(x) : x \in K \right\}.$$

- If K is a weakly compact subset of X, then $\overline{\mathrm{conv}}K$ is also weakly compact.

The following property holds for both topologies.

- If K is a weakly compact subset of X, then it is bounded. If K is a weakly* compact subset of a dual space X^*, then K is bounded.

The above is a simple consequence of the Banach Steinhaus Theorem, which in the case of continuous linear functionals takes the form:

Theorem 2.4 (Banach-Steinhaus Theorem) *Let X be a Banach space and A be a subset of X^*. The following statements are equivalent:*

(i) *A is bounded, that is, $\exists_{M>0} \forall_{x^* \in A} \|x^*\|_* \leq M$.*

(ii) *For each $x \in X$, the set $\{x^*(x) : x^* \in A\}$ is bounded.*

If X is an infinitely dimensional Banach space, then the closed unit ball B_X is never compact in norm topology. However, we have a very important fact concerning weak* topology.

- **(Alaoglu's Theorem)**. The closed unit ball B^* in X^* is always compact in the weak* topology.

Consequently,

- If X is reflexive, then the closed unit ball B in X is compact in the weak topology.

The following, very deep result states that the weak compactness is equivalent to the sequential weak compactness.

- **(Eberlein-Smulian Theorem)**. Let A be the subset of a Banach space X. The following statements are equivalent.

  (i) A is weakly compact.

(ii) Each sequence $\{x_n\} \subset A$ has a subsequence that converges weakly to a point of A.

This property does not hold in general for the weak* topologies. However, if the Banach space X is separable, then the relative weak* topology on the unit ball B^* in X^* is metrizable. Indeed, suppose $A = \{x_i\}_{i=1}^{\infty}$ is a countable, dense subset of $B \subset X$. Then, we define a metric d on B^* by putting for each $x^*, y^* \in B^*$

$$d(x^*, y^*) = \sum_{i=1}^{\infty} \frac{1}{2^i} |x^*(x_i) - y^*(x_i)| .$$

One can notice that the topology on B^* generated by the metric d coincides with the relative weak* topology on B^*. Then, for any sequence $\{x_n^*\}$ in B^*, we have

$$w^* - \lim_{n\to\infty} x_n^* = x^* \iff \lim_{n\to\infty} d(x_n^*, x^*) = 0.$$

It is easy to note that the same is true for any bounded set in X^*. Consequently, we have the following properties:

- Let X be a separable Banach space, and let A be a subset of X^*. The following statements are equivalent:

(i) A is weakly* compact.

(ii) Each sequence $\{x_n^*\} \subset A$ has a subsequence that converges weakly* to a point of A.

Below, we shall collect several known characterizations of reflexive Banach spaces.

Theorem 2.5 *Let X be a Banach space. The following statements are equivalent.*

(a) *X is reflexive.*

(b) *X^* is reflexive.*

(c) *(Alaoglu's Theorem). B_X is weakly compact.*

(d) *(Eberlein-Smulian Theorem). Any bounded sequence in X has a weakly convergent subsequence.*

(e) *(James's Theorem). For any $x^* \in X^*$, there exists $x \in B_X$ such that $x^*(x) = \|x^*\|_*$.*

(f) *(James's Theorem). For any bounded, closed and convex set $C \subset X$ and any $x^* \in X^*$, there exists $x \in C$ such that*

$$x^*(x) = \sup \{x^*(y) : y \in C\} .$$

(g) *(Smulian's Theorem).* *Any descending sequence $\{C_n\}$ of nonempty, bounded, closed and convex subsets of X has nonempty intersection; that is, $\bigcap_{n=1}^{\infty} C_n \neq \emptyset$.*

From the above, we immediately obtain that every finite dimensional Banach space is reflexive. If Y is a closed subspace of reflexive Banach space X, then Y is also reflexive. Using Smulian's or James's characterization, one can show that every uniformly convex space is reflexive. We shall prove it by showing that each functional $x^* \in X^*$ attains its norm on the unit sphere S_X. Indeed, suppose X is uniformly convex with the modulus of convexity δ_X. Take any functional $x^* \in X^*$. Without loss of generality, we can assume that $x^* \in S_{X^*}$. For any $n \in \mathbb{N}$, consider the *slice*

$$S\left(x^*, \delta_X\left(\frac{1}{n}\right)\right) = \left\{x \in B_X : x^*(x) \geq 1 - \delta_X\left(\frac{1}{n}\right)\right\}.$$

Then, for any two points $x, y \in S\left(x^*, \delta_X\left(\frac{1}{n}\right)\right)$, we have

$$1 - \delta_X\left(\frac{1}{n}\right) \leq x^*\left(\frac{x+y}{2}\right) \leq \|x^*\| \frac{\|x+y\|}{2} = \frac{\|x+y\|}{2}.$$

Since X is uniformly convex, we conclude that $\|x - y\| \leq \frac{1}{n}$, and, consequently, $\mathrm{diam}\left(S\left(x^*, \delta_X\left(\frac{1}{n}\right)\right)\right) \leq \frac{1}{n}$. Then, the intersection

$$\bigcap_{n \in \mathbb{N}} S\left(x^*, \delta_X\left(\frac{1}{n}\right)\right)$$

consists of exactly one point z, as it is a descending family of nonempty, closed sets with diameters converging to zero, and it must be the case that $x^*(z) = 1$.

Consequently, for each $1 < p < \infty$, the spaces $L_p[0,1]$ and ℓ_p are reflexive. Moreover, $L_p[0,1]^* = L_q[0,1]$ and $\ell_p^* = \ell_q$, where $\frac{1}{p} + \frac{1}{q} = 1$.

The space ℓ_1 is nonreflexive. To see it, consider the linear functional $x^* \in \ell_1^*$ defined for any $x = (x_1, x_2, \dots) \in \ell_1$ by

$$x^*(x) = \sum_{n=1}^{\infty} \left(1 - \frac{1}{n}\right) x_n.$$

It is easy to show that $\|x^*\| = 1$. However, x^* does not attain its norm on the unit ball B_{ℓ_1}. Also, the functional x^* generates the family of descending sequence $\{C_n\}_{n=1}^{\infty}$ of nonempty, bounded, closed and convex sets with empty intersection. Indeed, it is enough to consider *slices* defined as

$$C_n = \left\{x \in B_{\ell_1} : x^*(x) \geq 1 - \frac{1}{n}\right\}.$$

Since $c_0^* = \ell_1$ and $\ell_1^* = \ell_\infty$, we immediately get that c_0 and ℓ_∞ are nonreflexive. The space c is nonreflexive because it is isomorphic to c_0. The space

$L_1[0,1]$ contains an isometric copy of ℓ_1; hence, $L_1[0,1]$ is also nonreflexive. Moreover, $L_1[0,1]^* = L_\infty[0,1]$, so $L_\infty[0,1]$ is nonreflexive. Also, the space $C[0,1]$ is nonreflexive.

We say that a Banach space Y is *finitely representable* in a Banach space X if for each finite dimensional subspace Y_1 of Y and for each $\mu > 0$, there exist a subspace X_1 of X and an isomorphism $T : Y_1 \to X_1$, satisfying for any $y \in Y_1$ the following condition:

$$(1 - \mu) \|y\| \le \|Ty\| \le (1 + \mu) \|y\|.$$

A Banach space X is said to be *super-reflexive* if any Banach space Y that is finitely representable in X is reflexive.

We say that a Banach space X is *uniformly nonsquare* or has a *uniformly nonsquare norm* if $\epsilon_0(X) < 2$. The concept of uniformly nonsquare space is due to James.

Theorem 2.6 *Let X be a Banach space. The following statements are equivalent:*

(i) *X is super-reflexive.*

(ii) *(James, 1964). X admits an equivalent uniformly nonsquare norm.*

(iii) *(Enflo, 1972). X admits an equivalent uniformly convex norm.*

Obviously, every super-reflexive space X is also reflexive because X is finitely representable in X. However, there are examples of reflexive spaces that fail to be super-reflexive.

Now we can generalize an example 2.2 as follows. Let $(X, \|\cdot\|)$ be a Banach space that is not super-reflexive. Suppose that $\|\cdot\|_\circ$ is a strictly convex norm on X such that, for any $x \in X$,

$$\|x\|_\circ \le \|x\|.$$

Then, for any $\mu > 0$, the formula

$$\|\|x\|\|_\mu = \left(\|x\|^2 + \mu \|x\|_\circ^2 \right)^{\frac{1}{2}}$$

defines the norm in X, which is strictly convex and equivalent to the norm $\|\cdot\|$,

$$\|x\| \le \|\|x\|\|_\mu \le (1 + \mu)^{\frac{1}{2}} \|x\|.$$

However, $(X, \|\cdot\|)$ does not admit any equivalent uniformly nonsquare norm. Consequently,

$$\delta_{(X, \|\|\cdot\|\|_\mu)}(\epsilon) = \begin{cases} 0 & \text{if} \quad \epsilon \in [0, 2), \\ 1 & \text{if} \quad \epsilon = 2. \end{cases}$$

Recall that a sequence $\{x_n\}$ in a Banach space X is said to be a *Schauder basis* or a *basis* for X if, for every $x \in X$, there is a unique sequence $\{a_n\}$ of real numbers such that

$$x = \sum_{n=1}^{\infty} a_n x_n = \lim_{k \to \infty} \sum_{n=1}^{k} a_n x_n.$$

For instance, the sequence $\{e_n\}_{n=1}^{\infty}$, where

$$e_n = (\underbrace{0, \ldots, 0}_{n-1}, 1, 0, \ldots, 0, \ldots),$$

is the standard basis for c_0 and ℓ_p, $1 \leq p < \infty$. However, if we want to obtain a basis for the space c, we have to add to the above sequence one more element $e = (1, 1, \ldots, 1, \ldots)$. The space ℓ_∞ does not have Schauder basis because it is not separable.

The Schauder basis forms a linearly dense set; hence, we immediately obtain the following characterizations of weakly and weakly* convergent sequences.

- Let $\{x^n\} = \{(x_1^n, x_2^n, \ldots)\}$ be a bounded sequence in c_0 or ℓ_p, $1 < p < \infty$. Then, $\{x^n\}$ converges weakly to $x = (x_1, x_2, \ldots)$ if and only if, for each $i = 1, 2, \ldots$,

$$\lim_{n \to \infty} x_i^n = x_i.$$

- If $\{x^n\} = \{(x_1^n, x_2^n, \ldots)\}$ is a bounded sequence in $\ell_1 = c_0^*$, then $\{x^n\}$ converges weakly* to $x = (x_1, x_2, \ldots) \in \ell_1$ if and only if, for each $i = 1, 2, \ldots$,

$$\lim_{n \to \infty} x_i^n = x_i.$$

In other words, in the above-mentioned cases, weak convergence in c_0 and ℓ_p with $1 < p < \infty$ (or weak* convergence in $\ell_1 = c_0^*$) is equivalent to boundness and coordinate-wise convergence.

In the case of ℓ_1 space, we have the following important property.

- (**Schur's Lemma**) Let $\{x^n\}$ be a bounded sequence in ℓ_1. Then, $\{x^n\}$ is weakly convergent to $x \in \ell_1$ if and only if $\lim_{n \to \infty} \|x - x^n\| = 0$. Hence, the weak convergence is equivalent to the norm convergence.

It may happen that two different Banach spaces have the same dual space X^*. For instance, we have such a situation for the space ℓ_1, which is a dual space for both c_0 and c. Consequently, we have two different weak* topologies on ℓ_1, which, in fact, are not comparable. Later, we shall discuss this fact more precisely.

Chapter 3

Mean Lipschitz condition

In 2007, Goebel and Japón Pineda [33] introduced a class of **mean nonexpansive** mappings to obtain some new fixed point theorems of stability type. Their result initiated a new classification of Lipschitz mappings. Now, we shall present the origin of the condition discussed.

3.1 Nonexpansive and mean nonexpansive mappings in Banach spaces

Standard situation in the study of fixed point theory for nonexpansive mapping is the following: suppose C is a bounded closed and convex subset of a Banach space $(X, \|\cdot\|)$ and $S : C \to C$ is a nonexpansive mapping; that is, for all $x, y \in C$,

$$\|Sx - Sy\| \le \|x - y\|.$$

The above setting is natural. If we omit one of the above-mentioned assumptions on C, then, even in simple situations, nonexpansiveness does not guarantee that S has a fixed point in C. The reader can find appropriate examples on subsets of real line.

Let us recall a basic and very important property of nonexpansive mappings. In the proof, we shall use the Banach's Contraction Mapping Principle, which is discussed in detail in chapter 6.

Lemma 3.1 *Let C be a bounded closed and convex subset of a Banach space $(X, \|\cdot\|)$. If $S : C \to C$ is nonexpansive, then*

$$d(S) := \inf \{\|x - Sx\| : x \in C\} = 0.$$

Proof. Let us fix $z \in C$ and $\epsilon \in (0, 1)$. Let $S_\epsilon : C \to C$ be defined by

$$S_\epsilon x = \epsilon z + (1 - \epsilon) S x.$$

Let us notice that S_ϵ is a contraction because

$$\|S_\epsilon x - S_\epsilon y\| = (1 - \epsilon) \|Sx - Sy\| \le (1 - \epsilon) \|x - y\|$$

for $x, y \in C$. By the Banach's Contraction Mapping Principle, there exists a unique point $x_\epsilon \in C$ such that $S_\epsilon x_\epsilon = x_\epsilon$. Thus,

$$
\begin{aligned}
\|x_\epsilon - Sx_\epsilon\| &= \|\epsilon z + (1 - \epsilon)Sx_\epsilon - Sx_\epsilon\| \\
&= \epsilon \|z - Sx_\epsilon\| \\
&\leq \epsilon \, \mathrm{diam}(C).
\end{aligned}
$$

Letting $\epsilon \to 0$ we get the conclusion.

$\square$

As a consequence of the above lemma, we get the existence of *an approximate fixed point sequence for S* (a.f.p.s. for short), that is, a sequence $\{x_n\}$ in C such that

$$
\lim_{n \to \infty} \|x_n - Sx_n\| = 0.
$$

The question about the existence of a fixed point for S is equivalent to the question about whether the continuous function $\varphi : C \to \mathrm{R}$ given by $\varphi(x) = \|x - Sx\|$ attains its infimum. Obviously, it is true when C is compact. However, in view of the celebrated Schauder's Fixed Point Theorem, which guarantees that each continuous self-mapping F defined on a nonempty compact and convex set K has a fixed point, such a result is trivial. That is why we usually think about a noncompact set. Despite the fact that $d(S) = 0$, Fix(S) may be empty as seen in the following:

Example 3.1 *Let c_0 be the space of all sequences $x = (x_1, x_2, \dots)$ converging to zero with a standard sup norm,*

$$
\|x\| = \|(x_1, x_2, \dots)\| = \sup_{i=1,2,\dots} |x_i|.
$$

Let $S : B \to B$ be a mapping defined by

$$
Sx = S(x_1, x_2, \dots) = (1, x_1, x_2, \dots).
$$

It is easy to verify that S is an isometry, and so it is nonexpansive. However, the mapping S is fixed point free. Indeed, $Sx = x$ implies that $x_i = 1$ for $i = 1, 2, \dots$. But $x = (1, 1, \dots, 1, \dots) \notin c_0$. Thus, Fix$(S) = \emptyset$.

We say that C has *the fixed point property* for nonexpansive mappings (the f.p.p. for short) if each nonexpansive mapping $S : C \to C$ has at least one fixed point. If for a given space X, all bounded, closed and convex sets C in X have the f.p.p., then we say that X *has the fixed point property for nonexpansive mappings* or X *has the f.p.p.* for short. It is known that the f.p.p. for the set C depends deeply on the geometrical properties of the space X or on the set C itself. We shall discuss this in the chapter devoted to nonexpansive mappings. Now, we only mention that, among all such sets, we find, for example, all bounded closed and convex sets in a uniformly convex space X ([17] and

[43]), all weakly compact, convex sets having normal structure [48], all weak*
compact, convex sets in ℓ_1 considered as a dual of c_0 [47], and all bounded
closed and convex sets in uniformly nonsquare Banach spaces [28]. An excellent
overview of the fixed point theory for nonexpansive mappings can be found in
the monograph by Goebel and Kirk [36] or *Handbook of Metric Fixed Point
Theory* [49].

The mean Lipschitz condition has its roots in considerations devoted to the
possibility of the extension of the following Generalized Banach's Contraction
Mapping Principle to the class of nonexpansive mappings:

Theorem 3.1 *Let (M, ρ) be a complete metric space. Suppose $T : M \to M$
satisfies the following condition: there exists an integer p and a number $k \in
[0, 1)$ such that for all $x, y \in M$ we have*

$$\min \left\{ \rho \left(T^i x, T^i y \right) : i = 1, \ldots, p \right\} \leq k \rho \left(x, y \right).$$

Then, T has exactly one fixed point.

It is easy to observe that for every such mapping $d(T) = 0$. If $p = 1$, then
T is a contraction and consequently has a unique fixed point. Nevertheless,
if $p \geq 2$, then the above condition does not imply the continuity of T, which
makes it hard to prove that $\inf \left\{ \rho \left(x, Tx \right) : x \in M \right\}$ is achieved. The proofs
can be found in [3] and [62].

What if we use the counterpart of the above theorem in the case of nonexpansive mappings settings? Let us check it!

Suppose C is a bounded, closed and convex subset of a Banach space
$(X, \|\cdot\|)$ and that $T : C \to C$ satisfies for any $x, y \in C$ the above condition
with $k = 1$ and $p = 2$; that is,

$$\min \left\{ \|Tx - Ty\|, \left\| T^2 x - T^2 y \right\| \right\} \leq \|x - y\|.$$

Does $\inf \left\{ \|x - Tx\| \right\} = 0$?

The above condition is satisfied by all nonexpansive mappings. Nevertheless, it is also justified by all mappings that have nonexpansive square, and,
consequently, the answer is **NO!**

To illustrate this, let us consider the following example presented in [33]:

Example 3.2 *Let $(X, \|\cdot\|) = (\mathbb{R}, |\cdot|)$ and $C = [0, 1] \subset \mathbb{R}$. Let $\varphi : [1/4, 1/2) \to
[3/4, 1]$ be a one to one function with $\varphi([1/4, 1/2)) = [3/4, 1]$. Let us define a
mapping $T : [0, 1] \to [0, 1]$ by*

$$Tx = \begin{cases} x + 1/2 & \text{if } x \in [0, 1/4), \\ \varphi(x) & \text{if } x \in [1/4, 1/2), \\ x - 1/2 & \text{if } x \in [1/2, 3/4), \\ \varphi^{-1}(x) & \text{if } x \in [3/4, 1]. \end{cases}$$

*It is easy to notice that the function T is not continuous. Moreover, it does
not have an approximate fixed point sequence because $|x - Tx| \geq 1/4$ for each
$x \in [0, 1]$. However, $T^2 x = x$ for each $x \in [0, 1]$, and, consequently, T^2 is
nonexpansive.*

Immediately, we can ask what will happen if instead of **min** we shall consider **max**. Nothing interesting because the condition

$$\max\left\{\|Tx - Ty\|, \|T^2x - T^2y\|\right\} \leq \|x - y\|$$

is equivalent to the nonexpansiveness of T.

However, between **min** and **max**, we have a large collection of averages. The simplest one is

$$\frac{\|Tx - Ty\| + \|T^2x - T^2y\|}{2} \leq \|x - y\|. \qquad (3.1)$$

The above condition is satisfied by all nonexpansive mappings, but it does not exhaust all cases. To see it, consider the following example (see [42]):

Example 3.3 ([42]) *As a subset C, we take a closed unit ball B in the ℓ_1 space of all absolutely summable sequences, $x = (x_1, x_2, \ldots)$, with a metric inherited from the standard norm $\|x\| = \sum_{i=1}^{\infty} |x_i|$. Let $\tau : [-1, 1] \to [-1, 1]$ be the function given by*

$$\tau(t) = \begin{cases} 2t + 1, & \text{if } t \in [-1, -1/2], \\ 0, & \text{if } t \in [-1/2, 1/2], \\ 2t - 1, & \text{if } t \in [1/2, 1]. \end{cases}$$

It is easy to observe that, for all $s, t \in [-1, 1]$,

$$|\tau(s) - \tau(t)| \leq 2|s - t|,$$

and

$$|\tau(t)| \leq |t|.$$

Let us define a mapping $T : B \to B$ by

$$Tx = T(x_1, x_2, \ldots) = \left(\tau(x_2), \frac{2}{3}x_3, x_4, x_5, \ldots\right).$$

Then, for $i \geq 2$, we have

$$T^i x = \left(\tau\left(\frac{2}{3}x_{i+1}\right), \frac{2}{3}x_{i+2}, x_{i+3}, x_{i+4}, \ldots\right).$$

For each $x = (x_1, x_2, \ldots)$ and $y = (y_1, y_2, \ldots)$ in B we have

$$\begin{aligned} \|Tx - Ty\| &= |\tau(x_2) - \tau(y_2)| + \frac{2}{3}|x_3 - y_3| + \sum_{k=4}^{\infty}|x_k - y_k| \\ &\leq 2|x_2 - y_2| + \frac{2}{3}|x_3 - y_3| + \sum_{k=4}^{\infty}|x_k - y_k| \\ &\leq 2\|x - y\| \end{aligned}$$

and for $i \geq 2$

$$
\begin{aligned}
\left\| T^i x - T^i y \right\| &= \left| \tau\left(\frac{2}{3} x_{i+1}\right) - \tau\left(\frac{2}{3} y_{i+1}\right) \right| + \frac{2}{3} |x_{i+2} - y_{i+2}| \\
&\quad + \sum_{k=i+3}^{\infty} |x_k - y_k| \\
&\leq \frac{4}{3} |x_{i+1} - y_{i+1}| + \frac{2}{3} |x_{i+2} - y_{i+2}| + \sum_{k=i+3}^{\infty} |x_k - y_k| \\
&\leq \frac{4}{3} \|x - y\|.
\end{aligned}
$$

Let e_i *denote* i*-th vector of standard Schauder basis in* ℓ_1. *Then,*

$$
\left\| T e_2 - T \frac{3}{4} e_2 \right\| = \left| \tau(1) - \tau\left(\frac{3}{4}\right) \right| = 1 - \frac{1}{2} = \frac{1}{2} = 2 \left\| e_2 - \frac{3}{4} e_2 \right\|.
$$

For $i \geq 2$, *we have*

$$
\begin{aligned}
\left\| T^i e_{i+1} - T^i \frac{9}{10} e_{i+1} \right\| &= \left| \tau\left(\frac{2}{3}\right) - \tau\left(\frac{3}{5}\right) \right| = \frac{1}{3} - \frac{1}{5} = \frac{2}{15} \\
&= \frac{4}{3} \left\| e_{i+1} - \frac{9}{10} e_{i+1} \right\|.
\end{aligned}
$$

Thus, $k(T) = 2$ *and* $k(T^i) = 4/3$ *for* $i \geq 2$. *Despite of this, for all* $x, y \in B$, *we have*

$$
\begin{aligned}
\frac{1}{2} \|T x - T y\| + \frac{1}{2} \left\| T^2 x - T^2 y \right\| &\leq |x_2 - y_2| + |x_3 - y_3| + \frac{5}{6} |x_4 - y_4| \\
&\quad + \sum_{k=5}^{\infty} |x_k - y_k| \leq \|x - y\|.
\end{aligned}
$$

Hence, in the case of a closed unit ball in ℓ_1 space, the class of mappings that satisfies (3.1) is a proper extension of the class of nonexpansive ones.

Immediately, we can ask about a more general form of (3.1),

$$
\alpha_1 \|T x - T y\| + \alpha_2 \left\| T^2 x - T^2 y \right\| \leq \|x - y\|,
$$

where $\alpha_1 > 0$, $\alpha_2 > 0$ and $\alpha_1 + \alpha_2 = 1$. Why not consider more than two iterates of T? Let us introduce the following definitions.

In this text, we shall say that $\alpha = (\alpha_1, \ldots, \alpha_n)$ is a *multi-index*, if $n \geq 1$, $\alpha_1 > 0$, $\alpha_n > 0$, $\alpha_i \geq 0$, and $\sum_{i=1}^{n} \alpha_i = 1$. The number n will be called the *length* of the multi-index $\alpha = (\alpha_1, \ldots, \alpha_n)$.

Definition 3.1 (Goebel and Japón Pineda, 2007) *Fix a multi-index* $\alpha = (\alpha_1, \ldots, \alpha_n)$. *We say that a mapping* $T : C \to C$ *is* α-nonexpansive *if, for all* $x, y \in C$, *we have*

$$
\sum_{i=1}^{n} \alpha_i \left\| T^i x - T^i y \right\| \leq \|x - y\|.
$$

For $n = 1$, we obtain the classical definition of a nonexpansive mapping. If α is not specified, then we say that T is *mean nonexpansive*.

The importance of this theory is building up the intuition of the above class mappings. It could be observed in example 3.3 that such mappings may expand the distance between some points while sending them to such subset on which the second iterate contracts it back.

There is a nice characterization of mean nonexpansive mappings. Suppose that $T : C \to C$ is α-nonexpansive with $\alpha = (\alpha_1, \ldots, \alpha_n)$. Then, for all $x, y \in C$, we have

$$\sum_{i=1}^{n} \alpha_i \left\| T^i x - T^i y \right\| \le \| x - y \| .$$

Adding to both sides of the above inequality the term

$$\sum_{i=2}^{n} \left(\sum_{j=i}^{n} \alpha_j \right) \left\| T^{i-1} x - T^{i-1} y \right\|$$

we get

$$\sum_{i=1}^{n} \left(\sum_{j=i}^{n} \alpha_j \right) \left\| T^i x - T^i y \right\| \le \sum_{i=1}^{n} \left(\sum_{j=i}^{n} \alpha_j \right) \left\| T^{i-1} x - T^{i-1} y \right\| .$$

From the last inequality, we conclude that T is nonexpansive with respect to an equivalent metric d given for $x, y \in C$ by

$$d(x, y) = \| x - y \| + (\alpha_2 + \cdots + \alpha_n) \| Tx - Ty \| + \cdots + \alpha_n \left\| T^{n-1} x - T^{n-1} y \right\| .$$

Reassuming, a mapping $T : C \to C$ is α-nonexpansive with $\alpha = (\alpha_1, \ldots, \alpha_n)$ if and only if it is nonexpansive with respect to the equivalent metric d defined as above. In particular, every mean nonexpansive mapping is uniformly lipschitzian; however, the implication in opposite direction does not hold, as seen in the following:

Example 3.4 *Again, consider the space ℓ_1 with a standard norm and the mapping $T : B \to B$ given by*

$$Tx = T(x_1, x_2, \ldots) = (\tau(x_2), x_3, x_4, \ldots),$$

where τ is defined as in example 3.3. Then, T is uniformly lipschitzian with $k(T^i) = 2$ for any $i = 1, 2, \ldots$. In spite of this, T is not mean nonexpansive, for any α.

It is natural to ask about the behavior of consecutive Lipschitz constants for iterates T^m of T.

Suppose that $T : C \to C$ is α-nonexpansive with $\alpha = (\alpha_1, \alpha_2)$; that is,

$$\alpha_1 \left\| Tx - Ty \right\| + \alpha_2 \left\| T^2 x - T^2 y \right\| \leq \left\| x - y \right\|.$$

Standard reasoning leads us to the estimates

$$k\left(T \right) \leq \frac{1}{\alpha_1} \quad \text{and} \quad k\left(T^2 \right) \leq \min \left\{ \frac{1}{\alpha_1^2}, \frac{1}{\alpha_2} \right\}.$$

On the other hand, the mapping T is nonexpansive with respect to the metric d defined as

$$d\left(x, y \right) = \left\| x - y \right\| + \alpha_2 \left\| Tx - Ty \right\|.$$

The metric d is equivalent to the metric inherited by the norm $\left\| \cdot \right\|$, and for all $x, y \in C$, we have

$$\left\| x - y \right\| \leq d\left(x, y \right) \leq \left(1 + \frac{\alpha_2}{\alpha_1} \right) \left\| x - y \right\| = \frac{1}{\alpha_1} \left\| x - y \right\|.$$

Consequently, for all $m = 1, 2, \ldots,$

$$\left\| T^m x - T^m y \right\| \leq d\left(T^m x, T^m y \right) \leq d\left(x, y \right) \leq \frac{1}{\alpha_1} \left\| x - y \right\|.$$

Thus, $k\left(T^m \right) \leq \frac{1}{\alpha_1}$. In particular,

$$k\left(T^2 \right) \leq \min \left\{ \frac{1}{\alpha_1^2}, \frac{1}{\alpha_2}, \frac{1}{\alpha_1} \right\} = \min \left\{ \frac{1}{\alpha_1}, \frac{1}{\alpha_2} \right\}. \tag{3.2}$$

Now we find an interesting part of the story. It occurs that the above estimate is not sharp! To see it, first observe that, for all $x, y \in C$, we have

$$\alpha_1 \left\| T^2 x - T^2 y \right\| + \alpha_2 \left\| T^3 x - T^3 y \right\| \leq \left\| Tx - Ty \right\|.$$

Multiplying both sides of the above inequality by α_1 and adding $\alpha_2 \left\| T^2 x - T^2 y \right\|$ to both sides, we obtain

$$\left(\alpha_1^2 + \alpha_2 \right) \left\| T^2 x - T^2 y \right\| + \alpha_1 \alpha_2 \left\| T^3 x - T^3 y \right\|$$

$$\leq \alpha_1 \left\| Tx - Ty \right\| + \alpha_2 \left\| T^2 x - T^2 y \right\|$$

$$\leq \left\| x - y \right\|.$$

Consequently,

$$\left\| T^2 x - T^2 y \right\| \leq \frac{1}{\alpha_1^2 + \alpha_2} \left\| x - y \right\|.$$

Since $\alpha_1 + \alpha_2 = 1$, we get $\alpha_1^2 + \alpha_2 = \alpha_1 \left(1 - \alpha_2 \right) + \alpha_2 = 1 - \alpha_1 \alpha_2$. Thus,

$$k(T^2) \leq \frac{1}{1 - \alpha_1 \alpha_2}, \tag{3.3}$$

which is better than the estimate given by (3.2)! For instance, if $\alpha = (1/2, 1/2)$, then from (3.2), we get $k\left(T^2\right) \leq 2$, whereas using (3.3) we obtain $k(T^2) \leq 4/3$. Moreover, the construction presented in example 3.3 shows that this estimate is sharp!

The above observations come from [33]. Also, the authors posed the following problems (see [33]):

For any α, can we fully characterize the best Lipschitz constants of iterates of T? Is the class of mean nonexpansive mappings a proper extension of nonexpansive ones? Is this true regardless of the selection of a convex set C?

Later, we shall present answers to the above questions.

Let us come back to the general case of mean nonexpansive mappings. For each α-nonexpansive mapping $T : C \to C$,

$$\sum_{i=1}^{n} \alpha_i \left\| T^i x - T^i y \right\| \leq \|x - y\|, \tag{3.4}$$

we define the mapping $T_\alpha : C \to C$ by putting for each $x \in C$

$$T_\alpha x = \sum_{i=1}^{n} \alpha_i T^i x. \tag{3.5}$$

It is easy to observe that T_α is nonexpansive. Indeed, for all $x, y \in C$, we have

$$
\begin{aligned}
\|T_\alpha x - T_\alpha y\| &= \left\| \sum_{i=1}^{n} \alpha_i T^i x - \sum_{i=1}^{n} \alpha_i T^i y \right\| \\
&= \left\| \sum_{i=1}^{n} \alpha_i \left(T^i x - T^i y \right) \right\| \\
&\leq \sum_{i=1}^{n} \alpha_i \left\| T^i x - T^i y \right\| \\
&\leq \|x - y\|.
\end{aligned}
$$

Thus, in view of lemma 3.1,

$$d(T_\alpha) = \inf \left\{ \|x - T_\alpha x\| : x \in C \right\} = 0.$$

The above observations suggest that T_α may be treated as a "connection" between the class of mean nonexpansive mappings and the class of nonexpansive ones. Not surprisingly, this mapping plays a fundamental role in the metric fixed point theory for mean nonexpansive mappings.

However, the nonexpansiveness of T_α is much weaker than (3.4). The following example, presented in [42], shows that it does not imply the continuity of T.

Example 3.5 *Let $T : [0, 1] \to [0, 1]$ be defined as*

$$
Tx = \begin{cases} 1 & \text{if } x \in [0, 1/2], \\ 0 & \text{if } x \in (1/2, 1]. \end{cases}
$$

Then,

$$T^2 x = \begin{cases} 0 & \text{if } x \in [0, 1/2], \\ 1 & \text{if } x \in (1/2, 1]. \end{cases}$$

Thus, T and T^2 are discontinuous and

$$d(T) = \inf\left\{|x - Tx| : x \in [0, 1]\right\} = 1/2 > 0.$$

However, for $\alpha = (1/2, 1/2)$ and for each $x \in [0, 1]$, we have

$$T_\alpha x = \frac{1}{2} Tx + \frac{1}{2} T^2 x = \frac{1}{2}.$$

Hence, T_α is a constant and so nonexpansive mapping.

Moreover, even if T_α is nonexpansive and T is continuous, or even uniformly lipschitzian, it does not imply that (3.4) is satisfied and that $d(T) = 0$ as seen below.

Example 3.6 *(see [42]) Let $(X, \|\cdot\|)$ be an infinite dimensional Banach space. Since the work of Benyamini and Sternfeld [10], it is known that the unit sphere S is the lipschitzian retract of the closed unit ball B. It means that there exists a lipschitzian mapping $R : B \to S$ such that $Rx = x$ for all $x \in S$. Such retraction must have a sufficiently large Lipschitz constant. Standard evaluation is $k(R) \geq 3$. Define $T : B \to S$ as $T = -R$. Then, $T^2 = R$, and all the iterates of T satisfy $T^n = (-1)^n R$. Hence, all the powers of T have the same Lipschitz constant as R, $k(T^n) = k(R) \geq 3$. Since for any $\alpha = (\alpha_1, \ldots, \alpha_n)$*

$$\sum_{i=1}^{n} \alpha_i \left\| T^i x - T^i y \right\| = \|Rx - Ry\|, \tag{3.6}$$

T does not satisfy (3.4). However, for $\alpha = (1/2, 1/2)$

$$T_\alpha x = \frac{1}{2} Tx + \frac{1}{2} T^2 x = 0$$

for all $x \in B$. In spite of this, T does not have an approximate fixed point sequence. Indeed, for all $x \in B$, we have

$$2 = \left\| Tx - T^2 x \right\| \leq k(T) \|x - Tx\|,$$

which implies

$$\|x - Tx\| \geq \frac{2}{k(T)} > 0.$$

Thus, $d(T) > 0$.

The condition of mean nonexpansiveness can be treated as a perturbation of nonexpansiveness in the following sense: if α_1 is close to 1, then the perturbation is small, and for $\alpha_1 = 1$, we obtain nonexpansiveness of T, whereas if α_1 is close to 0, then the perturbation of nonexpansiveness is large, and for $\alpha_1 = 0$, we can even lose the continuity of T.

By analogy to the classical case of nonexpansive mappings, we say that C has the fixed point property for α-nonexpansive mappings if each α-nonexpansive mapping $T : C \to C$ has at least one fixed point.

Below we present some basic fixed point theorems for mean nonexpansive mappings, which can be treated as stability results of f.p.p. for nonexpansive mappings.

Theorem 3.2 (Goebel and Japón Pineda [33]) *If* $T : C \to C$ *is* (α_1, α_2)-*nonexpansive with* $\alpha_1 \geq \frac{1}{2}$, *then*

$$d(T) = \inf \{ \|x - Tx\| : x \in C \} = 0.$$

Proof. Since the mapping T_α is nonexpansive this implies that for any $\epsilon > 0$ there exists a point $x_\epsilon \in C$ such that

$$\|x_\epsilon - T_\alpha x_\epsilon\| = \|x_\epsilon - \alpha_1 T x_\epsilon - \alpha_2 T^2 x_\epsilon\| \leq \alpha_2 \epsilon.$$

By (α_1, α_2)-nonexpansiveness of T, we have

$$
\begin{aligned}
\alpha_1 \|Tx_\epsilon - T^2 x_\epsilon\| + \alpha_2 \|T^2 x_\epsilon - T^3 x_\epsilon\| &\leq \|x_\epsilon - Tx_\epsilon\| \\
&\leq \|x_\epsilon - T_\alpha x_\epsilon\| + \|T_\alpha x_\epsilon - Tx_\epsilon\| \\
&\leq \alpha_2 \epsilon + (1 - \alpha_1) \|Tx_\epsilon - T^2 x_\epsilon\|.
\end{aligned}
$$

Putting the last term on the right hand side to the left we obtain

$$(2\alpha_1 - 1) \|Tx_\epsilon - T^2 x_\epsilon\| + \alpha_2 \|T^2 x_\epsilon - T^3 x_\epsilon\| \leq \alpha_2 \epsilon.$$

Since $\alpha_1 \geq \frac{1}{2}$ the above inequality implies

$$\|T^2 x_\epsilon - T^3 x_\epsilon\| \leq \epsilon.$$

This means that $d(T) = 0$.

$\square$

If C has the f.p.p. for nonexpansive mappings, then we can repeat the above proof with $\epsilon = 0$ to obtain the following

Theorem 3.3 (Goebel and Japón Pineda [33]) *If* C *has the f.p.p. for nonexpansive mappings, then, for* $n = 2$, C *has the f.p.p. for* α-*nonexpansive mappings with all* $\alpha = (\alpha_1, \alpha_2)$ *such that* $\alpha_1 \geq \frac{1}{2}$.

Proof. Let $z \in C$ be fixed under T_α, that is, $T_\alpha z = z$. Then,

$$\alpha_1 \|Tz - T^2 z\| + \alpha_2 \|T^2 z - T^3 z\| \leq \|z - Tz\|$$

$$\leq \ \|T_\alpha z - Tz\|$$
$$= \ (1 - \alpha_1)\|T^2 z - Tz\|.$$

Thus,

$$(2\alpha_1 - 1)\|Tz - T^2 z\| + \alpha_2 \|T^2 z - T^3 z\| \leq 0.$$

If $2\alpha_1 - 1 > 0$, then $Tz = T^2 z$. If $2\alpha_1 - 1 = 0$, then $T^2 z$ is fixed under T.

$\square$

In a general case of multi-index $\alpha = (\alpha_1, \ldots, \alpha_n)$, we have the following

Theorem 3.4 (Goebel and Japón Pineda, [33]) *If* $T : C \to C$ *is* α-*nonexpansive for* $\alpha = (\alpha_1, \ldots, \alpha_n)$ *with* $n \geq 2$ *and* $\alpha_1 \geq 2^{\frac{1}{1-n}}$, *then* $d(T) = 0$.

Proof. Fix $\epsilon > 0$ and let $x \in C$ be such that $\|x - T_\alpha x\| < \epsilon$. Then,

$$
\begin{aligned}
\sum_{i=1}^{n} \alpha_i \|T^i x - T^{i+1} x\| &\leq \ \|x - Tx\| \\
&\leq \ \|x - T_\alpha x\| + \|T_\alpha x - Tx\| \\
&\leq \ \left\|\sum_{i=1}^{n} \alpha_i T^i x - \left(\sum_{i=1}^{n} \alpha_i\right) Tx\right\| + \epsilon \\
&= \ \left\|\sum_{i=2}^{n} \alpha_i \left(T^i x - Tx\right)\right\| + \epsilon \\
&\leq \ \sum_{i=2}^{n} \alpha_i \|Tx - T^i x\| + \epsilon.
\end{aligned}
$$

For any $i = 2, \ldots, n$, we have

$$
\alpha_i \|Tx - T^i x\| = \alpha_i \left\|\sum_{j=2}^{i} \left(T^{j-1} x - T^j x\right)\right\| \leq \alpha_i \left(\sum_{j=2}^{i} \|T^{j-1} x - T^j x\|\right).
$$

Thus,

$$
\begin{aligned}
\sum_{i=1}^{n} \alpha_i \|T^i x - T^{i+1} x\| &\leq \ \sum_{i=2}^{n} \left[\alpha_i \left(\sum_{j=2}^{i} \|T^{j-1} x - T^j x\|\right)\right] + \epsilon \\
&= \ \sum_{i=2}^{n} \left(\sum_{j=2}^{i} \alpha_i \|T^{j-1} x - T^j x\|\right) + \epsilon \\
&= \ \sum_{j=2}^{n} \left[\left(\sum_{i=j}^{n} \alpha_i\right) \|T^{j-1} x - T^j x\|\right] + \epsilon.
\end{aligned}
$$

This implies that

$$\alpha_n \left\| T^n x - T^{n+1} x \right\| \leq \sum_{j=2}^{n} \left[\left(\sum_{i=j}^{n} \alpha_i \right) - \alpha_{j-1} \right] \left\| T^{j-1} x - T^j x \right\| + \epsilon. \quad (3.7)$$

The first term on the right-hand side equals

$$\left[\left(\sum_{i=2}^{n} \alpha_i \right) - \alpha_1 \right] \left\| T x - T^2 x \right\| = (1 - 2\alpha_1) \left\| T x - T^2 x \right\|.$$

Further, for any $j = 3, \ldots, n$ we have

$$\left(\sum_{i=j}^{n} \alpha_i \right) - \alpha_{j-1} = 1 - 2\alpha_{j-1} - \left(\sum_{i=1}^{j-2} \alpha_i \right) \leq 1 - \alpha_1.$$

Using the above and the fact that $k(T^j) \leq 1/\alpha_1^j$, we obtain the evaluation

$$
\begin{aligned}
\alpha_n \left\| T^n x - T^{n+1} x \right\| &\leq (1 - 2\alpha_1) \left\| T x - T^2 x \right\| \\
&\quad + \sum_{j=3}^{n} (1 - \alpha_1) \left\| T^{j-1} x - T^j x \right\| + \epsilon \\
&\leq \left[1 - 2\alpha_1 + (1 - \alpha_1) \left(\sum_{j=3}^{n} \frac{1}{\alpha_1^{j-2}} \right) \right] \left\| T x - T^2 x \right\| + \epsilon \\
&= \left[1 - 2\alpha_1 + (1 - \alpha_1) \frac{\sum_{j=3}^{n} \alpha_1^{j-3}}{\alpha_1^{n-2}} \right] \left\| T x - T^2 x \right\| + \epsilon \\
&= \left(1 - 2\alpha_1 + \frac{1 - \alpha_1^{n-2}}{\alpha_1^{n-2}} \right) \left\| T x - T^2 x \right\| + \epsilon.
\end{aligned}
$$

It is easy to see that

$$1 - 2\alpha_1 + \frac{1 - \alpha_1^{n-2}}{\alpha_1^{n-2}} \leq 0$$

holds if

$$\alpha_1 \geq 2^{\frac{1}{1-n}}.$$

$\square$

If the mapping T_α has a fixed point $x = T_\alpha x$, then we can repeat the above proof with $\epsilon = 0$. As a consequence, we get that $z = T^n x$ is fixed point of T. We shall summarize the last observation in the following theorem.

Theorem 3.5 (Goebel and Japón Pineda, [33]) *If C has the f.p.p. for nonexpansive mappings, then for any multi-index $\alpha = (\alpha_1, \ldots, \alpha_n)$ with $n \geq 2$ and $\alpha_1 \geq 2^{\frac{1}{1-n}}$, C has the f.p.p. for α-nonexpansive mappings.*

In other words, if C has the f.p.p. for nonexpansive mappings, then C has the f.p.p. for each mapping $T : C \to C$, which is nonexpansive with respect to the metric d defined for all $x, y \in C$ as

$$d(x, y) = \|x - y\| + (\alpha_2 + \cdots + \alpha_n) \|Tx - Ty\| + \cdots + \alpha_n \|T^{n-1}x - T^{n-1}y\|$$

for some multi-index $\alpha = (\alpha_1, \ldots, \alpha_n)$ with $\alpha_1 \geq 2^{\frac{1}{1-n}}$.

The authors also noticed that this evaluation, which is based only on the value of initial index α_1, is not exact. In fact, if $\alpha = (\alpha_1, \alpha_2, \alpha_3)$ with $\alpha_1 \geq \alpha_2 \geq \alpha_3$, then formula (3.7) takes the form

$$\begin{aligned} \alpha_3 \left\|T^3 x - T^4 x\right\| &\leq (1 - 2\alpha_1) \left\|Tx - T^2 x\right\| + (\alpha_3 - \alpha_2) \left\|T^2 x - T^3 x\right\| + \epsilon \\ &\leq (1 - 2\alpha_1) \left\|Tx - T^2 x\right\| + \epsilon. \end{aligned}$$

If $\alpha_1 \geq 1/2$, then T has the minimal displacement zero. Consequently, we get the following:

Theorem 3.6 (Goebel and Japón Pineda) *If $T : C \to C$ is α-nonexpansive with a multi-index $\alpha = (\alpha_1, \alpha_2, \alpha_3)$ such that $\alpha_1 \geq \frac{1}{2}$ and $\alpha_1 \geq \alpha_2 \geq \alpha_3$, then*

$$d(T) = \inf \{\|x - Tx\| : x \in C\} = 0.$$

Now, we can connect theorems 3.6 and 3.4 (for $n = 3$) to obtain

Theorem 3.7 (Goebel and Japón Pineda) *If $T : C \to C$ is α-nonexpansive with a multi-index $\alpha = (\alpha_1, \alpha_2, \alpha_3)$ such that*

$$\alpha_1 \in \left[\frac{1}{2}, \frac{\sqrt{2}}{2}\right) \text{ and } \alpha_2 \geq \frac{1}{2} - \frac{1}{2}\alpha_1$$

or

$$\alpha_1 \geq \frac{\sqrt{2}}{2},$$

then

$$d(T) = \inf \{\|x - Tx\| : x \in C\} = 0.$$

Nevertheless, the following problem, posed by Goebel and Japón Pineda, is still open:

For $n = 2, 3, \ldots$ determine the set of all multi-indexes α of length n such that each α-nonexpansive mapping $T : C \to C$ has $d(T) = 0$.

3.2   General case

It seems natural to extend the idea presented by Goebel and Japón Pineda to the whole class of lipschitzian self-mappings defined on a metric space (M, ρ) (see [42]).

Definition 3.2 (Goebel, Japón Pineda, Sims) *Fix $k \geq 0$ and a multi-index $\alpha = (\alpha_1, \ldots, \alpha_n)$. A mapping $T : M \to M$ is said to be α-lipschitzian for the constant k if, for every $x, y \in M$, we have*

$$\sum_{i=1}^{n} \alpha_i \rho \left(T^i x, T^i y \right) \leq k \rho \left(x, y \right). \tag{3.8}$$

We referred to (3.8) as the *mean Lipschitz condition*. The smallest constant k for which (3.8) holds is called the *mean Lipschitz constant for T*, and it is denoted by $k(\alpha, T)$.

Observe that any α-lipschitzian mapping is also lipschitzian and

$$k(T) \leq \frac{k(\alpha, T)}{\alpha_1}.$$

Moreover, for $i = 1, \ldots, n$, provided $\alpha_i > 0$, we have

$$k(T^i) \leq \min \left[k(T)^i, \frac{k(\alpha, T)}{\alpha_i} \right] \leq \min \left[\left(\frac{k(\alpha, T)}{\alpha_1} \right)^i, \frac{k(\alpha, T)}{\alpha_i} \right]. \tag{3.9}$$

On the other hand, if T is lipschitzian, then for any multi-index α the mapping T is α-lipschitzian with

$$k(\alpha, T) \leq \sum_{i=1}^{n} \alpha_i k(T^i).$$

The class of all mappings that satisfies the *mean Lipschitz condition* with $\alpha = (\alpha_1, \ldots, \alpha_n)$ and $k \geq 0$ is denoted by $L(\alpha, k)$. Obviously, for $n = 1$, we get nothing more than the classical definition of k-lipschitzian mapping. Hence, in this special case, the mean Lipschitz condition determines the class $L(k)$.

If $k(\alpha, T) \leq 1$, then we say that T is α-*nonexpansive*, and if $k(\alpha, T) < 1$, then T is called α-*contraction*. We say that T is *mean nonexpansive* (*mean contraction*) if there exists a multi-index α such that T is α-nonexpansive (α-contraction). When the multi-index α and the constant k are not explicitly specified, we refer to such mapping as a *mean lipschitzian*.

For any multi-index $\alpha = (\alpha_1, \ldots, \alpha_n)$ and $k \geq 0$, each class $L(\alpha, k)$ contains all the lipschitzian mappings T such that

$$\sum_{i=1}^{n} \alpha_i k(T^i) \leq k.$$

Thus, using estimate $k(T^i) \leq k(T)^i$, we conclude that each class $L(\alpha, k)$ contains all the lipschitzian mappings T such that

$$\sum_{i=1}^{n} \alpha_i k(T)^i \leq k. \tag{3.10}$$

In particular, if $k \geq 1$, then each class $L(\alpha, k)$ contains all nonexpansive mappings; that is, $L(1) \subset L(\alpha, k)$.

If T is uniformly lipschitzian with $\sup \{k(T^i) : i = 1, 2, \ldots\} \leq k$, then for any α, T is α-lipschitzian with mean Lipschitz constant $k(\alpha, T) \leq k$. Hence, for any α, $L_u(k) \subset L(\alpha, k)$.

Suppose $T \in L(\alpha, k)$, where $\alpha = (\alpha_1, \ldots, \alpha_n)$ and $k > 0$. Consider a metric d defined for all $x, y \in M$ as

$$d(x, y) = \sum_{j=1}^{n} \left(\sum_{i=j}^{n} \alpha_i \right) \rho(T^{j-1}x, T^{j-1}y). \tag{3.11}$$

The metric d is equivalent to ρ and

$$\rho(x, y) \leq d(x, y) \leq b\rho(x, y)$$

with

$$b = 1 + \left(\sum_{i=2}^{n} \alpha_i \right) k_\rho(T) + \left(\sum_{i=3}^{n} \alpha_i \right) k_\rho(T^2) + \ldots + \alpha_n k_\rho(T^{n-1}).$$

Then,

$$
\begin{aligned}
d(Tx, Ty) &= \sum_{j=1}^{n} \left(\sum_{i=j}^{n} \alpha_i \right) \rho(T^j x, T^j y) \\
&= \sum_{i=1}^{n} \alpha_i \rho(T^i x, T^i y) + d(x, y) - \rho(x, y) \\
&\leq k\rho(x, y) + d(x, y) - \rho(x, y) \\
&= d(x, y) - (1 - k)\rho(x, y).
\end{aligned}
$$

If $k < 1$, then

$$d(Tx, Ty) \leq d(x, y) - \frac{1 - k}{b} d(x, y) = \frac{b - 1 + k}{b} d(x, y).$$

Thus,

$$k_d(T) \leq (b - 1 + k)/b < 1.$$

We shall summarize the above observation in the following theorem:

Theorem 3.8 (Goebel and Sims, [42]) *Let* (M, ρ) *be a metric space and* $T : M \to M$ *be an* α-*contraction. Then, there exists a metric* d *equivalent to* ρ *such that* T *is a contraction with respect to* d.

If (M, ρ) is a complete metric space, then (M, d) is also a complete metric space and, by the Banach's Contraction Mapping Principle, the mapping T has a unique fixed point. Thus, we get the following:

Theorem 3.9 *Let* (M, ρ) *be a complete metric space. If* $T : M \to M$ *is an* α-*contraction, then* T *has the unique fixed point in* M.

If a mapping T is α-nonexpansive ($k = 1$) with respect to ρ, then T is nonexpansive with respect to d. Consequently, T is uniformly lipschitzian and the inequality

$$\rho(T^n x, T^n y) \leq d(T^n x, T^n y) \leq d(x, y) \leq b\rho(x, y)$$

holds for all $x, y \in M$ and $n = 1, 2, \ldots$.
If $k(\alpha, T) > 1$, then, using estimate $\rho(x, y) \leq d(x, y)$, we get

$$d(Tx, Ty) \leq d(x, y) - (1 - k)\rho(x, y) \leq d(x, y) - (1 - k)d(x, y) = kd(x, y).$$

In view of the above remarks, we obtain the following:

Theorem 3.10 *Any* α-*lipschitzian mapping* $T : M \to M$ *with* $k(\alpha, T) \geq 1$ *is lipschitzian with respect to the equivalent metric* d *defined by (3.11) with* $k_d(T) \leq k(\alpha, T)$.

Chapter 4

On the Lipschitz constants for iterates of mean lipschitzian mappings

The mean Lipschitz condition involves only a finite numbers of iterates. In spite of this, as we shall see later, this condition has a serious influence not only on behavior of the sequence of Lipschitz constants for consecutive iterates but also on its asymptotic behavior expressed in terms of constants $k_0(T)$ and $k_\infty(T)$.

We start our investigation with a simple case of mean lipschitzian mappings with $\alpha = (\alpha_1, \alpha_2)$. We will obtain some results that will be useful to predict some analogous in general case of multi-index α of length n.

4.1 A bound for Lipschitz constants of iterates

We have already seen that, for any mapping $T \in L((1/2, 1/2), 1)$, we have $k(T) \leq 2$ and $k(T^2) \leq 4/3$ and that this estimate is optimal.

In the following theorem, we present sharp evaluation of Lipschitz constants for all iterates of (α_1, α_2)-nonexpansive mappings.

Theorem 4.1 *Let $T : M \to M$ be (α_1, α_2)-nonexpansive. Then, for $n \geq 0$, we have*

$$k(T^n) \leq \frac{1 + \alpha_2}{1 + (-1)^n \alpha_2^{n+1}}.$$

Proof. It is easy to verify that the above bound is satisfied for $n = 0$ and $n = 1$. For $n = 2$, we have

$$
\begin{aligned}
\alpha_2 \rho(T^2 x, T^2 y) &\leq \rho(x, y) - \alpha_1 \rho(Tx, Ty) \\
&\leq \rho(x, y) - \alpha_1^2 \rho(T^2 x, T^2 y).
\end{aligned}
$$

Thus,

$$
\begin{aligned}
\rho(T^2 x, T^2 y) &\leq \frac{1}{\alpha_1^2 + \alpha_2} \rho(x, y) \\
&= \frac{1 + \alpha_2}{1 + \alpha_2^3} \rho(x, y).
\end{aligned}
$$

For $n \geq 3$, we have

$$
\begin{aligned}
\rho(T^n x, T^n y) &\leq \frac{1}{\alpha_2} \rho(T^{n-2} x, T^{n-2} y) - \frac{\alpha_1}{\alpha_2} \rho(T^{n-1} x, T^{n-1} y) \\
&= \frac{1 - (-\alpha_2)^{n-2}}{\alpha_2 + (-\alpha_2)^n} \left[\alpha_1 \rho(T^{n-2} x, T^{n-2} y) + \alpha_2 \rho(T^{n-1} x, T^{n-1} y) \right] \\
&\quad + \frac{\alpha_2 + (-\alpha_2)^{n-2}}{\alpha_2 + (-\alpha_2)^n} \rho(T^{n-2} x, T^{n-2} y) \\
&\quad + \frac{-1 + (-\alpha_2)^n}{\alpha_2 + (-\alpha_2)^n} \rho(T^{n-1} x, T^{n-1} y) \\
&\leq \frac{1 - (-\alpha_2)^{n-2}}{\alpha_2 + (-\alpha_2)^n} \rho(T^{n-3} x, T^{n-3} y) \\
&\quad + \frac{\alpha_2 + (-\alpha_2)^{n-2}}{\alpha_2 + (-\alpha_2)^n} \rho(T^{n-2} x, T^{n-2} y) \\
&\quad + \frac{-1 + (-\alpha_2)^n}{\alpha_2 + (-\alpha_2)^n} \alpha_1 \rho(T^n x, T^n y).
\end{aligned}
$$

By putting the last term on the right-hand side to the left and multiplying both sides by

$$
\alpha_2 + (-\alpha_2)^n,
$$

we obtain

$$
\begin{aligned}
\left[1 - (-\alpha_2)^{n+1} \right] \rho(T^n x, T^n y) &\leq \left[1 - (-\alpha_2)^{n-2} \right] \rho(T^{n-3} x, T^{n-3} y) \\
&\quad + \alpha_2 \left[1 - (-\alpha_2)^{n-3} \right] \rho(T^{n-2} x, T^{n-2} y).
\end{aligned}
$$

If $n = 3$, then

$$
\rho(T^3 x, T^3 y) \leq \frac{1 + \alpha_2}{1 - \alpha_2^4} \rho(x, y).
$$

For $n \geq 4$, we have

$$
\begin{aligned}
&\left[1 - (-\alpha_2)^{n+1} \right] \rho(T^n x, T^n y) \\
&\leq \left[1 - (-\alpha_2)^{n-2} \right] \rho(T^{n-3} x, T^{n-3} y) + \alpha_2 \left[1 - (-\alpha_2)^{n-3} \right] \rho(T^{n-2} x, T^{n-2} y) \\
&= \left[1 - (-\alpha_2)^{n-3} \right] \left(\alpha_1 \rho(T^{n-3} x, T^{n-3} y) + \alpha_2 \rho(T^{n-2} x, T^{n-2} y) \right) \\
&\quad + \left[\alpha_2 + (-\alpha_2)^{n-3} \right] \rho(T^{n-3} x, T^{n-3} y) \\
&\leq \left[1 - (-\alpha_2)^{n-3} \right] \rho(T^{n-4} x, T^{n-4} y) + \alpha_2 \left[1 - (-\alpha_2)^{n-4} \right] \rho(T^{n-3} x, T^{n-3} y) \\
&= \left[1 - (-\alpha_2)^{n-4} \right] \left(\alpha_1 \rho(T^{n-4} x, T^{n-4} y) + \alpha_2 \rho(T^{n-3} x, T^{n-3} y) \right) \\
&\quad + \left[\alpha_2 + (-\alpha_2)^{n-4} \right] \rho(T^{n-4} x, T^{n-4} y) \\
&\leq \left[1 - (-\alpha_2)^{n-4} \right] \rho(T^{n-5} x, T^{n-5} y) + \alpha_2 \left[1 - (-\alpha_2)^{n-5} \right] \rho(T^{n-4} x, T^{n-4} y) \\
&\leq \cdots \\
&\leq (1 + \alpha_2) \rho(x, y) + (\alpha_2 - \alpha_2) \rho(Tx, Ty) \\
&= (1 + \alpha_2) \rho(x, y).
\end{aligned}
$$

Thus,

$$\rho(T^n x, T^n y) \le \frac{1 + \alpha_2}{1 - (-\alpha_2)^{n+1}} \rho(x, y).$$

□

The following example shows that the estimate presented in the above theorem is optimal. Moreover, for arbitrary $\alpha = (\alpha_1, \alpha_2)$, we construct a mapping T, which is (α_1, α_2)-nonexpansive, and all its iterates T^n have the biggest possible Lipschitz constant; that is,

$$k(T^n) = \frac{1 + \alpha_2}{1 - (-\alpha_2)^{n+1}}$$

for $n = 1, 2, \ldots$.

Example 4.1 *As a metric space M, we consider the space ℓ_1 with a metric inherited from the standard norm, $\|x\| = \|(x_1, x_2, \ldots)\| = \sum_{i=1}^{\infty} |x_i|$. Let $\alpha = (\alpha_1, \alpha_2)$ be as above and $T : \ell_1 \to \ell_1$ be a mapping defined for every $x = (x_1, x_2, \ldots) \in \ell_1$ by*

$$Tx = \left(\frac{1 - (-\alpha_2)^1}{1 - (-\alpha_2)^2} x_2, \frac{1 - (-\alpha_2)^2}{1 - (-\alpha_2)^3} x_3, \ldots, \frac{1 - (-\alpha_2)^j}{1 - (-\alpha_2)^{j+1}} x_{j+1}, \ldots \right).$$

Then,

$$T^2 x = \left(\frac{1 - (-\alpha_2)^1}{1 - (-\alpha_2)^3} x_3, \frac{1 - (-\alpha_2)^2}{1 - (-\alpha_2)^4} x_4, \frac{1 - (-\alpha_2)^3}{1 - (-\alpha_2)^5} x_5, \ldots \right).$$

In general, for $n = 2, 3, \ldots$, we have

$$T^n x = \left(\frac{1 - (-\alpha_2)^1}{1 - (-\alpha_2)^{n+1}} x_{n+1}, \frac{1 - (-\alpha_2)^2}{1 - (-\alpha_2)^{n+2}} x_{n+2}, \frac{1 - (-\alpha_2)^3}{1 - (-\alpha_2)^{n+3}} x_{n+3}, \ldots \right).$$

Let us observe that, for any $x = (x_1, x_2, \ldots) \in \ell_1$, we get

$$\alpha_1 \|Tx\| + \alpha_2 \|T^2 x\|$$

$$= \alpha_1 \sum_{i=1}^{\infty} \frac{1 - (-\alpha_2)^i}{1 - (-\alpha_2)^{i+1}} |x_{i+1}| + \alpha_2 \sum_{i=1}^{\infty} \frac{1 - (-\alpha_2)^i}{1 - (-\alpha_2)^{i+2}} |x_{i+2}|$$

$$= |x_2| + \alpha_1 \sum_{i=1}^{\infty} \frac{1 - (-\alpha_2)^{i+1}}{1 - (-\alpha_2)^{i+2}} |x_{i+2}| + \alpha_2 \sum_{i=1}^{\infty} \frac{1 - (-\alpha_2)^i}{1 - (-\alpha_2)^{i+2}} |x_{i+2}|$$

$$= |x_2| + \sum_{i=1}^{\infty} \left(\alpha_1 \frac{1 - (-\alpha_2)^{i+1}}{1 - (-\alpha_2)^{i+2}} + \alpha_2 \frac{1 - (-\alpha_2)^i}{1 - (-\alpha_2)^{i+2}} \right) |x_{i+2}|$$

$$= |x_2| + \sum_{i=1}^{\infty} |x_{i+2}|$$

$$\le \|x\|.$$

Since T is linear, the above means that it is α-nonexpansive. In particular, this implies that T is well defined; that is, for any $x \in \ell_1$, $Tx \in \ell_1$. In view of the previous theorem,

$$k(T^n) \leq \frac{1 + \alpha_2}{1 + (-1)^n \alpha_2^{n+1}}$$

for every $n = 1, 2, \ldots$. If e_n denotes the n-th vector of standard basis in ℓ_1, then

$$\|T^n e_{n+1}\| = \left\| \left(\frac{1 - (-\alpha_2)^1}{1 - (-\alpha_2)^{n+1}}, 0, 0, \ldots \right) \right\| = \frac{1 + \alpha_2}{1 + (-1)^n \alpha_2^{n+1}}.$$

Consequently,

$$k(T^n) = \frac{1 + \alpha_2}{1 + (-1)^n \alpha_2^{n+1}},$$

as we desired.

For the convenience of the reader, we present several samples of graph of the sequence

$$b_n = \frac{1 + \alpha_2}{1 + (-1)^n \alpha_2^{n+1}} \tag{4.1}$$

as well as vectors of approximate values of its initial terms. For clarity, we connect points of graph with lines.

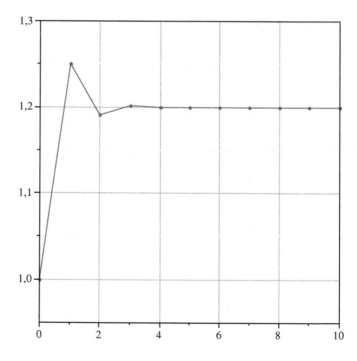

FIGURE 4.1: A graph of $\{b_n\}$ with $\alpha = (\frac{4}{5}, \frac{1}{5})$.

$[b_0, \ldots, b_{15}] \approx [1., \ 1.250000000, \ 1.190476190, \ 1.201923077, \ 1.199616123,$
$1.200076805, \ 1.199984640, \ 1.200003072, \ 1.199999386, \ 1.200000123, \ 1.199999975,$
$1.200000005, \ 1.199999999, \ 1.200000000, \ 1.200000000, \ 1.200000000]$

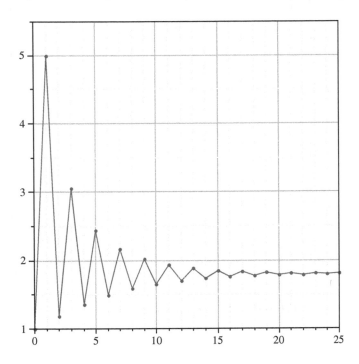

FIGURE 4.2: A graph of $\{b_n\}$ with $\alpha = (\frac{1}{5}, \frac{4}{5})$.

$[b_0, \ldots, b_{100}] \approx [1., \ 5., \ 1.190476190, \ 3.048780488, \ 1.355748373, \ 2.439500390,$
$1.487953528, \ 2.162869245, \ 1.586996884, \ 2.016522449, \ 1.657612197, \ 1.932822555,$
$1.706200628, \ 1.882806715, \ 1.738820686, \ 1.852132907, \ 1.760360212, \ 1.833020767,$
$1.774427800, \ 1.820994639, \ 1.783549658, \ 1.813380386, \ 1.789437028, \ 1.808540591,$
$1.793225386, \ 1.805456658, \ 1.795658364, \ 1.803488454, \ 1.797218938, \ 1.802231054,$
$1.798219130, \ 1.801427238, \ 1.798859837, \ 1.800913171, \ 1.799270129, \ 1.800584323,$
$1.799532815, \ 1.800373923, \ 1.799700973, \ 1.800239293, \ 1.799808612, \ 1.800153140,$
$1.799877507, \ 1.800098007, \ 1.799921602, \ 1.800062723, \ 1.799949825, \ 1.800040142,$
$1.799967887, \ 1.800025691, \ 1.799979448, \ 1.800016442, \ 1.799986847, \ 1.800010523,$
$1.799991582, \ 1.800006735, \ 1.799994612, \ 1.800004310, \ 1.799996552, \ 1.800002758,$
$1.799997793, \ 1.800001765, \ 1.799998588, \ 1.800001130, \ 1.799999096, \ 1.800000723,$
$1.799999422, \ 1.800000463, \ 1.799999630, \ 1.800000296, \ 1.799999763, \ 1.800000190,$
$1.799999848, \ 1.800000121, \ 1.799999903, \ 1.800000078, \ 1.799999938, \ 1.800000050,$
$1.799999960, \ 1.800000032, \ 1.799999975, \ 1.800000020, \ 1.799999984, \ 1.800000013,$
$1.799999990, \ 1.800000008, \ 1.799999993, \ 1.800000005, \ 1.799999996, \ 1.800000003,$
$1.799999997, \ 1.800000002, \ 1.799999998, \ 1.800000001, \ 1.799999999, \ 1.800000001,$
$1.799999999, \ 1.800000001, \ 1.800000000, \ 1.800000000, \ 1.800000000]$

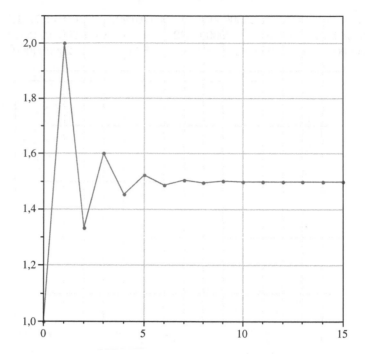

FIGURE 4.3: A graph of $\{b_n\}$ with $\alpha = (\frac{1}{2}, \frac{1}{2})$.

$[b_0, \ldots, b_{40}] \approx [1., 2., 1.333333333, 1.600000000, 1.454545455,$
$1.523809524, 1.488372093, 1.505882353, 1.497076023, 1.501466276,$
$1.499267936, 1.500366300, 1.499816917, 1.500091558, 1.499954225,$
$1.500022889, 1.499988556, 1.500005722, 1.499997139, 1.500001431,$
$1.499999285, 1.500000358, 1.499999821, 1.500000089, 1.499999955,$
$1.500000022, 1.499999989, 1.500000006, 1.499999997, 1.500000001,$
$1.499999999, 1.500000000, 1.500000000, 1.500000000, 1.500000000,$
$1.500000000, 1.500000000, 1.500000000, 1.500000000, 1.500000000,$
$1.500000000]$

It is not easy to guess the shape of the sequence $\{b_n\}_{n=0}^{\infty}$ in general case of $k > 0$. Indeed, as we shall see later, for any $T \in L((\alpha_1, \alpha_2), k)$, we have

$$k(T^n) \leq b_n,$$

where

$$b_n = k^n \frac{2^{n+1}\sqrt{\Delta}}{\left(\alpha_1 + \sqrt{\Delta}\right)^{n+1} - \left(\alpha_1 - \sqrt{\Delta}\right)^{n+1}},$$

with $\Delta = \alpha_1^2 + 4\alpha_2 k$.

Despite being complicated and sophisticated, the above-mentioned boundary is sharp! Nevertheless, this is still only the case of multi-index α of length $n = 2$. It may suggest that, in general, we should find a recurrence formula for the sequence $\{b_n\}_{n=0}^{\infty}$, provided it exists!

Let us start with some particular cases of $\{b_n\}_{n=0}^{\infty}$ given by (4.1).

For $\alpha = \left(\frac{1}{2}, \frac{1}{2}\right)$, we get

$$b_0 = 1, b_1 = 2, b_2 = \frac{4}{3}, b_3 = \frac{8}{5}, b_4 = \frac{16}{11}, b_5 = \frac{32}{21}, b_6 = \frac{64}{43}, b_7 = \frac{128}{85},$$

$$b_8 = \frac{256}{171}, b_9 = \frac{512}{341}, b_{10} = \frac{1024}{683}, b_{11} = \frac{2048}{1365}, b_{12} = \frac{4096}{2731}, b_{13} = \frac{8192}{5461},$$

$$b_{14} = \frac{16384}{10923}, b_{15} = \frac{32768}{21845}, b_{16} = \frac{65536}{43691}, b_{17} = \frac{131072}{87381}, b_{18} = \frac{262144}{174763},$$

$$b_{19} = \frac{524288}{349525}, b_{20} = \frac{1048576}{699051}, \ldots.$$

It is clear that the numerator of b_n equals 2^n. However, it is also easy to observe that the numerator (resp. denominator) of b_n equals 2 times numerator (denominator) of b_{n-2} plus numerator (denominator) of b_{n-1}; that is,

$$b_0 = 1 = \frac{1}{1}, b_1 = 2 = \frac{2}{1},$$

and

$$b_2 = \frac{2 \cdot 1 + 2}{2 \cdot 1 + 1} = \frac{4}{3}, b_3 = \frac{2 \cdot 2 + 4}{2 \cdot 1 + 3} = \frac{8}{5}, b_4 = \frac{2 \cdot 4 + 8}{2 \cdot 3 + 5} = \frac{16}{11},$$

$$b_5 = \frac{2 \cdot 8 + 16}{2 \cdot 5 + 11} = \frac{32}{21}, b_6 = \frac{2 \cdot 16 + 32}{2 \cdot 11 + 21} = \frac{64}{43}, \ldots.$$

For $\alpha = \left(\frac{1}{5}, \frac{4}{5}\right)$, we have

$$b_0 = 1, b_1 = 5, b_2 = \frac{25}{21}, b_3 = \frac{125}{41}, b_4 = \frac{625}{461}, b_5 = \frac{3125}{1281}, b_6 = \frac{15625}{10501},$$

$$b_7 = \frac{78125}{36121}, b_8 = \frac{390625}{246141}, b_9 = \frac{1953125}{968561}, b_{10} = \frac{9765625}{5891381}, \ldots.$$

Obviously, the numerator of b_n equals 5^n or equivalently 5 times the numerator of b_{n-2} plus 4 times the numerator of b_{n-1}, but the denominator of b_n equals 20 times the denominator of b_{n-2} plus the denominator of b_{n-1}, that is,

$$b_0 = 1 = \frac{1}{1}, b_1 = 5 = \frac{5}{1},$$

and

$$b_2 = \frac{5 \cdot 1 + 4 \cdot 5}{20 \cdot 1 + 1} = \frac{25}{21}, b_3 = \frac{5 \cdot 5 + 4 \cdot 25}{20 \cdot 1 + 21} = \frac{125}{41},$$

$$b_4 = \frac{5 \cdot 25 + 4 \cdot 125}{20 \cdot 21 + 41} = \frac{625}{461}, b_5 = \frac{5 \cdot 125 + 4 \cdot 625}{20 \cdot 41 + 461} = \frac{3125}{1281}, \ldots.$$

Following the method described above, the reader can find nice formulas for some other multi-indexes, for example,

$$\alpha = \left(\frac{4}{5}, \frac{1}{5}\right), \quad \alpha = \left(\frac{1}{3}, \frac{2}{3}\right), \quad \alpha = \left(\frac{2}{3}, \frac{1}{3}\right), \quad \text{and} \quad \alpha = \left(\frac{1}{4}, \frac{3}{4}\right).$$

However, the above procedure does not lead us to the recurrence formula in the general case of $\alpha = (\alpha_1, \alpha_2)$, especially if α has irrational coefficients. Nevertheless, the general recurrence formula exists and the reader can verify that the sequence $\{b_n\}_{n=0}^{\infty}$ defined by (4.1) satisfies the following relation: $b_0 = 1, b_1 = \frac{1}{\alpha_1}$, and

$$b_{n+2} = \frac{1}{\alpha_1 b_{n+1}^{-1} + \alpha_2 b_n^{-1}}.$$

It turns out that, in opposite to an explicit formula of $\{b_n\}_{n=0}^{\infty}$, its recurrence formula has a very easy and natural extension for $k > 0$, as can be observed in the following theorem.

Theorem 4.2 *Let $T : M \to M$ be (α_1, α_2)-lipschitzian mapping for the constant $k > 0$. Then, for $n \geq 0$, we have*

$$k(T^n) \leq b_n,$$

where

$$b_0 = 1, \quad b_1 = \frac{k}{\alpha_1} \quad \text{and} \quad b_{n+2} = \frac{k}{\alpha_1 b_{n+1}^{-1} + \alpha_2 b_n^{-1}}.$$

Proof. It is clear that our formula is correct for $n = 0$ and $n = 1$. For $n = 2$, by definition of T, we have

$$\alpha_2 \rho(T^2 x, T^2 y) \leq k\rho(x, y) - \alpha_1 \rho(Tx, Ty)$$
$$\leq k\rho(x, y) - \alpha_1 \frac{1}{b_1} \rho(T^2 x, T^2 y).$$

Thus,

$$\rho(T^2 x, T^2 y) \leq \frac{k}{\alpha_1 b_1^{-1} + \alpha_2 b_0^{-1}} \rho(x, y)$$
$$= b_2 \rho(x, y).$$

For $n \geq 3$, by definition of T, we can write

$$\alpha_2 \rho(T^n x, T^n y) \leq k\rho(T^{n-2} x, T^{n-2} y) - \alpha_1 \rho(T^{n-1} x, T^{n-1} y)$$
$$= k\rho(T^{n-2} x, T^{n-2} y) + \frac{b_{n-2}}{b_{n-3}} \alpha_2 \rho(T^{n-1} x, T^{n-1} y)$$
$$+ \left(-\alpha_1 - \frac{b_{n-2}}{b_{n-3}} \alpha_2\right) \rho(T^{n-1} x, T^{n-1} y)$$

$$\leq \quad k\rho(T^{n-2}x, T^{n-2}y) + \frac{b_{n-2}}{b_{n-3}}\alpha_2\rho(T^{n-1}x, T^{n-1}y)$$

$$+\frac{\alpha_1}{k}\left(-\alpha_1 - \frac{b_{n-2}}{b_{n-3}}\alpha_2\right)\rho(T^n x, T^n y).$$

Moving the last term from the right-hand side to the left and dividing both sides by b_{n-2}, we obtain

$$\frac{\alpha_1\left(\alpha_1 b_{n-3} + \alpha_2 b_{n-2}\right) + \alpha_2 k b_{n-3}}{k b_{n-3} b_{n-2}}\rho(T^n x, T^n y)$$

$$\leq k\frac{1}{b_{n-2}}\rho(T^{n-2}x, T^{n-2}y) + \alpha_2\frac{1}{b_{n-3}}\rho(T^{n-1}x, T^{n-1}y).$$

It is easy to verify that the term on the left-hand side equals

$$\alpha_1\frac{1}{b_{n-1}} + \alpha_2\frac{1}{b_{n-2}} = \frac{k}{b_n}.$$

Hence, we can write

$$\frac{k}{b_n}\rho(T^n x, T^n y) \leq k\frac{1}{b_{n-2}}\rho(T^{n-2}x, T^{n-2}y) + \alpha_2\frac{1}{b_{n-3}}\rho(T^{n-1}x, T^{n-1}y). \quad (4.2)$$

For $n = 3$, we have

$$\frac{k}{b_3}\rho(T^3 x, T^3 y) \quad \leq \quad k\frac{1}{b_1}\rho(Tx, Ty) + \alpha_2\frac{1}{b_0}\rho(T^2 x, T^2 y)$$

$$= \quad \frac{1}{b_0}\left(\alpha_1\rho(Tx, Ty) + \alpha_2\rho(T^2 x, T^2 y)\right)$$

$$+ \left(k\frac{1}{b_1} - \alpha_1\frac{1}{b_0}\right)\rho(Tx, Ty)$$

$$\leq \quad k\rho(x, y).$$

Thus,

$$\rho(T^3 x, T^3 y) \leq b_3\rho(x, y).$$

For $n \geq 4$, starting from the right-hand side of inequality (4.2), we repeat the following procedure from $i = 1$ to $n - 3$,

$$k\frac{1}{b_{n-i-1}}\rho(T^{n-i-1}x, T^{n-i-1}y) + \alpha_2\frac{1}{b_{n-i-2}}\rho(T^{n-i}x, T^{n-i}y)$$

$$= \frac{1}{b_{n-i-2}}\left(\alpha_1\rho(T^{n-i-1}x, T^{n-i-1}y) + \alpha_2\rho(T^{n-i}x, T^{n-i}y)\right)$$

$$+ \left(k\frac{1}{b_{n-i-1}} - \alpha_1\frac{1}{b_{n-i-2}}\right)\rho(T^{n-i-1}x, T^{n-i-1}y)$$

$$\leq k\frac{1}{b_{n-i-2}}\rho(T^{n-i-2}x, T^{n-i-2}y) + \alpha_2\frac{1}{b_{n-i-3}}\rho(T^{n-i-1}x, T^{n-i-1}y)$$

and after $(n-3)$ steps (or 1 step, if $n = 4$), we obtain

$$
\begin{aligned}
\frac{k}{b_n}\rho(T^n x, T^n y) &\leq k\frac{1}{b_1}\rho(Tx, Ty) + \alpha_2\frac{1}{b_0}\rho(T^2 x, T^2 y) \\
&= \frac{1}{b_0}\left(\alpha_1\rho(Tx, Ty) + \alpha_2\rho(T^2 x, T^2 y)\right) \\
&\quad + \left(k\frac{1}{b_1} - \alpha_1\frac{1}{b_0}\right)\rho(Tx, Ty) \\
&\leq k\rho(x, y).
\end{aligned}
$$

Thus,

$$
\rho(T^n x, T^n y) \leq b_n\rho(x, y).
$$

$\square$

To see that the above estimation is optimal, we consider the following:

Example 4.2 (*Left shift with special coefficients*) *Let $\{b_n\}_{n\geq 0}$ be as in theorem 4.2. Let $T : \ell_1 \to \ell_1$ be a linear mapping given for every $x = (x_1, x_2, \dots) \in \ell_1$ by*

$$
Tx = \left(\frac{b_1}{b_0}x_2, \frac{b_2}{b_1}x_3, \dots, \frac{b_j}{b_{j-1}}x_{j+1}, \dots\right).
$$

Then,

$$
T^2 x = \left(\frac{b_2}{b_0}x_3, \frac{b_3}{b_1}x_4, \frac{b_4}{b_2}x_5, \dots\right)
$$

and, in general,

$$
T^n x = \left(\frac{b_n}{b_0}x_{n+1}, \frac{b_{n+1}}{b_1}x_{n+2}, \frac{b_{n+2}}{b_2}x_{n+3}, \dots\right)
$$

for $n \geq 2$.

We claim that mapping T is (α_1, α_2)-lipschitzian for the constant $k > 0$. Indeed, for each $x \in \ell_1$, we have

$$
\begin{aligned}
\alpha_1\|Tx\| + \alpha_2\|T^2 x\| &= \alpha_1\sum_{i=1}^{\infty}\frac{b_i}{b_{i-1}}|x_{i+1}| + \alpha_2\sum_{i=1}^{\infty}\frac{b_{i+1}}{b_{i-1}}|x_{i+2}| \\
&= k|x_2| + \alpha_1\sum_{i=1}^{\infty}\frac{b_{i+1}}{b_i}|x_{i+2}| + \alpha_2\sum_{i=1}^{\infty}\frac{b_{i+1}}{b_{i-1}}|x_{i+2}| \\
&= k|x_2| + \sum_{i=1}^{\infty}\left(\alpha_1\frac{b_{i+1}}{b_i} + \alpha_2\frac{b_{i+1}}{b_{i-1}}\right)|x_{i+2}| \\
&= k|x_2| + k\sum_{i=1}^{\infty}|x_{i+2}| \\
&\leq k\|x\|.
\end{aligned}
$$

If e_n denotes the n-th vector of the standard Schauder basis in ℓ_1, then

$$\|T^n e_{n+1}\| = b_n,$$

and by theorem (4.2), we get $k(T^n) = b_n$.

Let us observe that, for $k = 1$, the above mapping coincide with those described in example 4.1.

Now, using standard methods, we give an explicit formula for the sequence $\{b_n\}_{n=0}^{\infty}$:

Lemma 4.1 *Let* $\{b_n\}_{n=0}^{\infty}$ *be as in theorem 4.2. Then, for* $n \geq 0$, *we have*

$$b_n = k^n \frac{2^{n+1}\sqrt{\Delta}}{\left(\alpha_1 + \sqrt{\Delta}\right)^{n+1} - \left(\alpha_1 - \sqrt{\Delta}\right)^{n+1}}$$

with $\Delta = \alpha_1^2 + 4\alpha_2 k$.

Proof. If we define $\{d_n\}_{n\geq 0}$ by $d_n = \frac{1}{b_n}$, then we obtain the following relation:

$$d_0 = 1, \; d_1 = \frac{\alpha_1}{k}, \text{ and } d_{n+2} = \frac{\alpha_1}{k}d_{n+1} + \frac{\alpha_2}{k}d_n, \; n \geq 0.$$

The solutions of the characteristic equation

$$\lambda^2 - \frac{\alpha_1}{k}\lambda - \frac{\alpha_2}{k} = 0$$

are

$$\lambda_1 = \frac{\alpha_1 + \sqrt{\Delta}}{2k} \text{ and } \lambda_2 = \frac{\alpha_1 - \sqrt{\Delta}}{2k},$$

with

$$\Delta := \alpha_1^2 + 4\alpha_2 k.$$

Hence, the general solution has the form

$$d_n = C_1\left(\frac{\alpha_1 + \sqrt{\Delta}}{2k}\right)^n + C_2\left(\frac{\alpha_1 - \sqrt{\Delta}}{2k}\right)^n, \; n \geq 0.$$

Using initial condition, we solve a linear system

$$\begin{cases} C_1 + C_2 = 1 \\ C_1\frac{\alpha_1+\sqrt{\Delta}}{2k} + C_2\frac{\alpha_1-\sqrt{\Delta}}{2k} = \frac{\alpha_1}{k} \end{cases}$$

and get

$$\begin{cases} C_1 = \frac{\alpha_1+\sqrt{\Delta}}{2\sqrt{\Delta}} \\ C_2 = -\frac{\alpha_1-\sqrt{\Delta}}{2\sqrt{\Delta}}. \end{cases}$$

Thus,

$$d_n = \frac{\left(\alpha_1 + \sqrt{\Delta}\right)^{n+1} - \left(\alpha_1 - \sqrt{\Delta}\right)^{n+1}}{2^{n+1} k^n \sqrt{\Delta}},$$

and, finally, for $n \geq 0$,

$$b_n = k^n \frac{2^{n+1} \sqrt{\Delta}}{\left(\alpha_1 + \sqrt{\Delta}\right)^{n+1} - \left(\alpha_1 - \sqrt{\Delta}\right)^{n+1}}.$$

$\square$

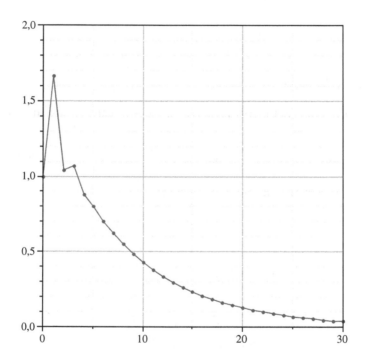

FIGURE 4.4: A graph of $\{b_n\}$, with $k = \frac{5}{6}$ and $\alpha = (\frac{1}{2}, \frac{1}{2})$.

$[b_0, \ldots, b_{50}] \approx [1.000 * 10^0, 1.667 * 10^0, 1.042 * 10^0, 1.068 * 10^0, 8.790 * 10^{-1}, 8.037 * 10^{-1}, 6.998 * 10^{-1}, 6.235 * 10^{-1}, 5.495 * 10^{-1}, 4.868 * 10^{-1}, 4.302 * 10^{-1}, 3.806 * 10^{-1}, 3.366 * 10^{-1}, 2.977 * 10^{-1}, 2.633 * 10^{-1}, 2.329 * 10^{-1}, 2.060 * 10^{-1}, 1.822 * 10^{-1}, 1.611 * 10^{-1}, 1.425 * 10^{-1}, 1.260 * 10^{-1}, 1.115 * 10^{-1}, 9.858 * 10^{-2}, 8.719 * 10^{-2}, 7.711 * 10^{-2}, 6.820 * 10^{-2}, 6.032 * 10^{-2}, 5.335 * 10^{-2}, 4.718 * 10^{-2}, 4.173 * 10^{-2}, 3.691 * 10^{-2}, 3.264 * 10^{-2}, 2.887 * 10^{-2}, 2.553 * 10^{-2}, 2.258 * 10^{-2}, 1.997 * 10^{-2}, 1.767 * 10^{-2}, 1.562 * 10^{-2}, 1.382 * 10^{-2}, 1.222 * 10^{-2}, 1.081 * 10^{-2}, 9.560 * 10^{-3}, 8.455 * 10^{-3}, 7.478 * 10^{-3}, 6.614 * 10^{-3}, 5.850 * 10^{-3}, 5.174 * 10^{-3}, 4.576 * 10^{-3}, 4.047 * 10^{-3}, 3.579 * 10^{-3}, 3.166 * 10^{-3}]$

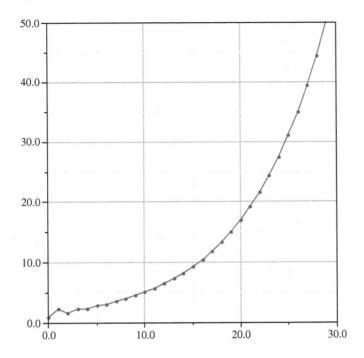

FIGURE 4.5: A graph of $\{b_n\}$, with $k = \frac{6}{5}$ and $\alpha = (\frac{1}{2}, \frac{1}{2})$.

$[b_0, \ldots, b_{30}] \approx [1., 2.4, 1.694117647, 2.383448276, 2.376618911,$
$2.856034433, 3.113237586, 3.574913812, 3.993769417, 4.527302101,$
$5.092599155, 5.752009524, 6.482707796, 7.314646544, 8.248314569,$
$9.304168566, 10.49338151, 11.83566936, 13.34901637, 15.05624926,$
$16.98159380, 19.15328259, 21.60261536, 24.36521932, 27.48108358,$
$30.99542723, 34.95918267, 39.42983712, 44.47220460, 50.15940250,$
$56.57388973]$

Corollary 4.1 *Let $T : M \to M$ be (α_1, α_2)-lipschitzian mapping for the constant $k > 0$. Then, for $n \geq 0$, we have*

$$k(T^n) \leq k^n \frac{2^{n+1}\sqrt{\Delta}}{\left(\alpha_1 + \sqrt{\Delta}\right)^{n+1} - \left(\alpha_1 - \sqrt{\Delta}\right)^{n+1}}, \qquad (4.3)$$

with $\Delta = \alpha_1^2 + 4\alpha_2 k$; see figures 4.4 and 4.5.

Let us observe that, for $k = 1$, we get

$$\Delta = \alpha_1^2 + 4\alpha_2 = (1 - \alpha_2)^2 + 4\alpha_2 = 1 + 2\alpha_2 + \alpha_2^2 = (1 + \alpha_2)^2.$$

Consequently, using evaluation (4.3), for any (α_1, α_2)-nonexpansive mapping $T : M \to M$, we get

$$
\begin{aligned}
k(T^n) &\leq \frac{2^{n+1}(1+\alpha_2)}{(\alpha_1+\alpha_2+1)^{n+1} - (\alpha_1 - 1 - \alpha_2)^{n+1}} \\
&= \frac{2^{n+1}(1+\alpha_2)}{2^{n+1} - (-2\alpha_2)^{n+1}} \\
&= \frac{2^{n+1}(1+\alpha_2)}{2^{n+1} + (-1)^n 2^{n+1} \alpha_2^{n+1}} \\
&= \frac{1+\alpha_2}{1 + (-1)^n \alpha_2^{n+1}}
\end{aligned}
$$

as in theorem 4.1.

Let us pass now to the general case of multi-index α of length n. The recurrence formula presented in theorem 4.2 may suggest the following conjecture in the general case: let $T : M \to M$ be $(\alpha_1, \ldots, \alpha_n)$-lipschitzian for the constant k, let $\{b_m\}$ be defined as

$$
b_m = \begin{cases}
1 & \text{for} \quad m = 0, \\[2mm]
\dfrac{k}{\sum\limits_{j=1}^{m} \alpha_j b_{m-j}^{-1}} & \text{for} \quad m = 1, \ldots, n, \\[6mm]
\dfrac{k}{\sum\limits_{i=1}^{n} \alpha_i b_{m-i}^{-1}} & \text{for} \quad m = n+1, n+2, \ldots;
\end{cases}
$$

under the above settings we have $k(T^m) \leq b_m$.

Actually, the above conjecture holds, and it was proved by Pérez García and the present author in [64].

To prove this result, we shall need the following lemma:

Lemma 4.2 *Let $T : M \to M$ be an $(\alpha_1, \ldots, \alpha_n)$-lipschitzian mapping for the constant $k > 0$. Let $b_0 = 1$, and for $m = 1, \ldots, n$, we set*

$$
b_m = \frac{k}{\sum\limits_{j=1}^{m} \alpha_j b_{m-j}^{-1}}.
$$

Then, $k(T^m) \leq b_m$ for $m = 1, \ldots, n$.

Proof. Obviously, $k(T) \leq k/\alpha_1 = b_1$.

Suppose that, for some m, with $1 \leq m \leq n$, we have already proved that

$$
k(T) \leq b_1, \ldots, k(T^{m-1}) \leq b_{m-1}.
$$

Since $\rho(T^m x, T^m y) \leq k(T^{m-1})\rho(Tx, Ty) \leq b_{m-1}\rho(Tx, Ty)$, this implies that

$$
\alpha_1 b_{m-1}^{-1} \rho(T^m x, T^m y) \leq \alpha_1 \rho(Tx, Ty).
$$

Further, $\rho(T^m x, T^m y) \leq k(T^{m-2})\rho(T^2 x, T^2 y) \leq b_{m-2}\rho(T^2 x, T^2 y)$, which implies that

$$\alpha_2 b_{m-2}^{-1} \rho(T^m x, T^m y) \leq \alpha_2 \rho(T^2 x, T^2 y).$$

Continuing this process, we obtain

$$\alpha_j b_{m-j}^{-1} \rho(T^m x, T^m y) \leq \alpha_j \rho(T^j x, T^j y)$$

for $j = 1, \ldots, m$.

Using the above and our assumption that T is $(\alpha_1, \ldots, \alpha_n)$-lipschitzian for $k > 0$, we get

$$\left(\alpha_1 b_{m-1}^{-1} + \cdots + \alpha_m b_0^{-1}\right) \rho(T^m x, T^m y) \leq \sum_{j=1}^{m} \alpha_j \rho(T^j x, T^j y) \leq k\rho(x, y).$$

Thus, $k(T^m) \leq b_m$.

$\square$

Let us note that the last lemma is valid even for the case $\alpha_n = 0$. Actually, we only used the fact that $\alpha_1 > 0$. We apply this observation in the proof of the following theorem:

Theorem 4.3 (Pérez García and Piasecki, [64]) *Let* $T : M \to M$ *be an* $(\alpha_1, \ldots, \alpha_n)$-*lipschitzian mapping for the constant* $k > 0$. *Then for* $m \geq 0$ *we have*

$$k(T^m) \leq b_m,$$

where b_m *is defined as follows:*

$$b_m = \begin{cases} 1 & \text{for} \quad m = 0, \\[2mm] \dfrac{k}{\sum_{j=1}^{m} \alpha_j b_{m-j}^{-1}} & \text{for} \quad m = 1, \ldots, n, \\[4mm] \dfrac{k}{\sum_{i=1}^{n} \alpha_i b_{m-i}^{-1}} & \text{for} \quad m = n+1, n+2, \ldots. \end{cases}$$

Proof. We have already proved that for $m = 1, \ldots, n$, $k(T^m) \leq b_m$. Now consider $m = n + j$, with $j \geq 1$. Since $T : M \to M$ is $(\alpha_1, \ldots, \alpha_n)$-lipschitzian, it is also $(\alpha_1, \ldots, \alpha_n, \alpha_{n+1}, \ldots, \alpha_m)$-lipschitzian, provided $\alpha_{n+1} = \alpha_{n+2} = \cdots = \alpha_m = 0$. Applying lemma 4.2 to this mapping, we obtain bounds for $k(T), \ldots, k(T^n), k(T^{n+1}), \ldots, k(T^m)$, which are precisely our desired bounds.

$\square$

To prove that the above bound is sharp, it is enough to consider a "copy" of example 4.2.

Example 4.3 *Fix $\alpha = (\alpha_1, \ldots, \alpha_n)$ and $k > 0$. Let $\{b_m\}_{m \geq 0}$ be the sequence defined as in theorem 4.3. Let us consider the space $M = \ell_1$ and the linear mapping $T : \ell_1 \to \ell_1$ given for every $x = (x_1, x_2, \ldots) \in \ell_1$ by*

$$Tx = \left(\frac{b_1}{b_0} x_2, \frac{b_2}{b_1} x_3, \ldots, \frac{b_j}{b_{j-1}} x_{j+1}, \ldots \right).$$

Then,

$$\alpha_1 Tx = \alpha_1 \left(\frac{b_1}{b_0} x_2, \frac{b_2}{b_1} x_3, \frac{b_3}{b_2} x_4, \ldots \right),$$

$$\alpha_2 T^2 x = \alpha_2 \left(\frac{b_2}{b_0} x_3, \frac{b_3}{b_1} x_4, \frac{b_4}{b_2} x_5, \ldots \right),$$

$$\vdots \qquad \qquad \vdots$$

$$\alpha_n T^n x = \alpha_n \left(\frac{b_n}{b_0} x_{n+1}, \frac{b_{n+1}}{b_1} x_{n+2}, \frac{b_{n+2}}{b_2} x_{n+3}, \ldots \right).$$

Reordering the elements of $\sum_{j=1}^{n} \alpha_j \|T^j x\|$ and using the definition of $\{b_m\}_{m \geq 0}$, we have that the term that contains $|x_2|$ equals

$$b_1(\alpha_1 b_0^{-1})|x_2| = b_1 \left(\frac{\alpha_1 b_0^{-1}}{k} \right) (k|x_2|) = b_1(b_1^{-1})k|x_2| = k|x_2|.$$

Similarly, the term that contains $|x_3|$ is

$$b_2(\alpha_1 b_1^{-1} + \alpha_2 b_0^{-1})|x_3| = b_2 \left(\frac{\alpha_1 b_1^{-1} + \alpha_2 b_0^{-1}}{k} \right) (k|x_3|) = b_2(b_2^{-1})k|x_3| = k|x_3|.$$

After doing this n times, we get that the term that contains $|x_{n+1}|$ equals $k|x_{n+1}|$. Now, the term that contains $|x_{n+1+j}|$ with $j \geq 1$ is

$$b_{n+j} \left(\sum_{i=1}^{n} \alpha_i b_{n+j-i}^{-1} \right) |x_{n+1+j}|$$

$$= b_{n+j} \left(\frac{\sum_{i=1}^{n} \alpha_i b_{n+j-i}^{-1}}{k} \right) (k|x_{n+1+j}|)$$

$$= b_{n+j}(b_{n+j}^{-1})k|x_{n+1+j}|$$

$$= k|x_{n+1+j}|.$$

Thus,

$$\sum_{j=1}^{n} \alpha_j \|T^j x\| = k \sum_{j=2}^{\infty} |x_j| \leq k\|x\|, \tag{4.4}$$

and, consequently, T is an $(\alpha_1, \ldots, \alpha_n)$-lipschitzian mapping for the constant k. This in particular implies that T is well defined; that is, for any $x \in \ell_1$, $Tx \in \ell_1$. In view of theorem 4.3, we have $k(T^m) \leq b_m$. If $\{e_j\}_{j=1}^{\infty}$ is the standard basis of ℓ_1, then we have $T^m(e_{m+1}) = b_m b_0^{-1} e_1 = b_m e_1$, which proves that the bounds b_m are sharp.

To show a behavior of the sequence $\{b_m\}_{m=0}^{\infty}$, with $k = 1$ and $n \geq 3$, we present some samples of its graph as well as vectors of approximate values of its initial terms:

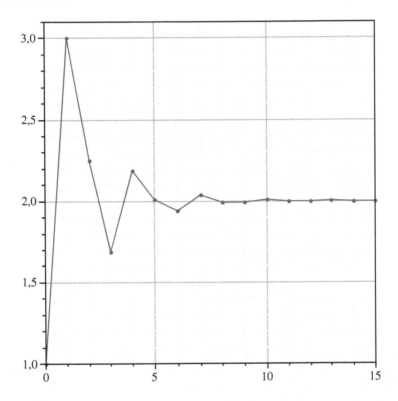

FIGURE 4.6: A graph of $\{b_m\}$, with $k = 1$ and $\alpha = (\frac{1}{3}, \frac{1}{3}, \frac{1}{3})$.

$[b_0, \ldots, b_{50}] \approx [1., 3., 2.250000000, 1.687500000, 2.189189189, 2.008264463, 1.938829787, 2.040111940, 1.994831256, 1.990393366, 2.008196164, 1.997778329, 1.998762623, 2.001568038, 1.999368374, 1.999898952, 2.000278016, 1.999848378, 2.000008430, 2.000044925, 1.999967241, 2.000006865, 2.000006343, 1.999993483, 2.000002230, 2.000000685, 1.999998800, 2.000000572, 2.000000019, 1.999999797, 2.000000129, 1.999999982, 1.999999969, 2.000000027, 1.999999992, 1.999999996, 2.000000005, 1.999999998, 2.000000000, 2.000000001, 2.000000000, 2.000000000, 2.000000000, 2.000000000, 2.000000000, 2.000000000, 2.000000000, 2.000000000, 2.000000000, 2.000000000, 2.000000000]$

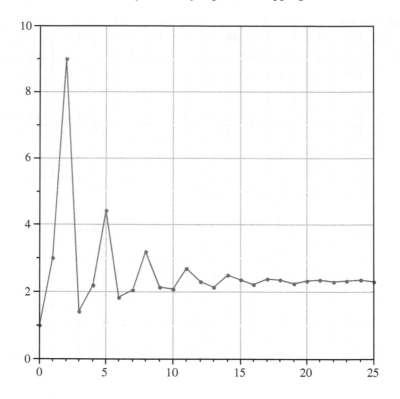

FIGURE 4.7: A graph of $\{b_m\}$, with $k = 1$ and $\alpha = (\frac{1}{3}, 0, \frac{2}{3})$.

$[b_1, \ldots, b_{100}]$ $\approx [1., \quad 3., \quad 9., \quad 1.421052632, \quad 2.189189189, \quad 4.418181818,$ $1.836272040, 2.057384760, 3.195811008, 2.139689097, 2.084106872, 2.713357943,$ $2.301915805, 2.151980866, 2.496292593, 2.363255025, 2.218079368, 2.396111347,$ $2.374106553, 2.267758689, 2.351742660, 2.366604812, 2.299776996, 2.334161787,$ $2.355690725, 2.318117632, 2.328789115, 2.346654729, 2.327552566, 2.328376786,$ $2.340530279, 2.331862450, 2.329537516, 2.336854515, 2.333524100, 2.330864863,$ $2.334854547, 2.333967414, 2.331898129, 2.333868242, 2.333934356, 2.332576477,$ $2.333437495, 2.333768712, 2.332973753, 2.333282894, 2.333606750, 2.333184714,$ $2.333250166, 2.333487876, 2.333285759, 2.333262030, 2.333412590, 2.333328034,$ $2.333284031, 2.333369735, 2.333341935, 2.333303332, 2.333347600, 2.333343823,$ $2.333316829, 2.333337343, 2.333341663, 2.333325107, 2.333333264, 2.333338864,$ $2.333329692, 2.333332074, 2.333336600, 2.333331995, 2.333332048, 2.333335083,$ $2.333333024, 2.333332373, 2.333334180, 2.333333409, 2.333332719, 2.333333693,$ $2.333333504, 2.333332980, 2.333333455, 2.333333488, 2.333333149, 2.333333353,$ $2.333333443, 2.333333247, 2.333333318, 2.333333401, 2.333333298, 2.333333311,$ $2.333333371, 2.333333323, 2.333333315, 2.333333353, 2.333333333, 2.333333321,$ $2.333333342, 2.333333336, 2.333333326, 2.333333337]$

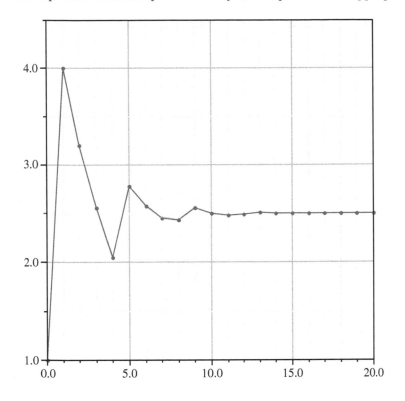

FIGURE 4.8: A graph of $\{b_m\}$, with $k = 1$ and $\alpha = \left(\frac{1}{4}, \frac{1}{4}, \frac{1}{4}, \frac{1}{4}\right)$.

$[b_0, \ldots, b_{53}] \approx [1., 4., 3.200000000, 2.560000000,$
$2.048000000, 2.775067751, 2.577721838, 2.458214554,$
$2.434020427, 2.554387333, 2.504594409, 2.486956285,$
$2.494247041, 2.509772290, 2.498860985, 2.497431729,$
$2.500064393, 2.501522973, 2.499469111, 2.499621188,$
$2.500169153, 2.500195344, 2.499863657, 2.499962313,$
$2.500047609, 2.500017225, 2.499972699, 2.499999961,$
$2.500009373, 2.499999815, 2.499995462, 2.500001153,$
$2.500001451, 2.499999470, 2.499999384, 2.500000364,$
$2.500000167, 2.499999846, 2.499999940, 2.500000080,$
$2.500000008, 2.499999969, 2.499999999, 2.500000014,$
$2.499999998, 2.499999995, 2.500000001, 2.500000002,$
$2.499999999, 2.499999999, 2.500000000, 2.500000000,$
$2.500000000, 2.500000000]$

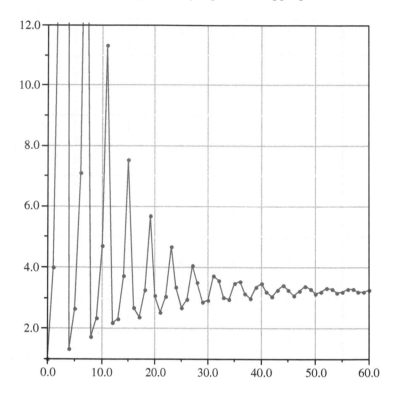

FIGURE 4.9: A graph of $\{b_m\}$, with $k = 1$ and $\alpha = \left(\frac{1}{4}, 0, 0, \frac{3}{4}\right)$.

$[b_0, \ldots, b_{200}] \approx [1., \quad 4., \quad 16., \quad 64., \quad 1.326424870, \quad 2.659740260, \quad 7.098786828,$
$21.30559168, 1.732610707, 2.345912569, 4.712086964, 11.33053647, 2.198106354,$
$2.307128288, 3.737968954, 7.514613324, 2.670431675, 2.388360419, 3.275273222,$
$5.677461873, 3.077991090, 2.530077792, 3.050643232, 4.671779165, 3.364984081,$
$2.697394453, 2.953932044, 4.078779738, 3.518939677, 2.864588948, 2.931077843,$
$3.715105219, 3.566012964, 3.012738075, 2.951075057, 3.489263571, 3.546510815,$
$3.130529285, 2.993981754, 3.350690894, 3.495440976, 3.214422884, 3.046208045,$
$3.269002834, 3.435940569, 3.267080674, 3.098578280, 3.224663003, 3.380567490,$
$3.294731996, 3.145393978, 3.204473522, 3.334754185, 3.304647211, 3.183750766,$
$3.199267583, 3.299817976, 3.303438577, 3.212852204, 3.202652957, 3.274978154,$
$3.296277198, 3.233310046, 3.210262597, 3.258555879, 3.286765230, 3.246510167,$
$3.219248395, 3.248639291, 3.277150095, 3.254116310, 3.227895133, 3.243428296,$
$3.268654077, 3.257738615, 3.235304625, 3.241393556, 3.261796050, 3.258752027,$
$3.241134786, 3.241328860, 3.256655058, 3.258227532, 3.245391137, 3.242343475,$
$3.253065331, 3.256935446, 3.248269535, 3.243822962, 3.250749805, 3.255386830,$
$3.250045940, 3.245376471, 3.249404804, 3.253889260, 3.251005918, 3.246782005,$
$3.248748707, 3.252602597, 3.251404941, 3.247936506, 3.248545619, 3.251587403,$
$3.251450554, 3.248814306, 3.248612786, 3.250843238, 3.251298704, 3.249435049,$
$3.248818313, 3.250336770, 3.251058167, 3.249840677, 3.249073844, 3.250020946,$

3.250798800, 3.250080155, 3.249325363, 3.249847023, 3.250560803, 3.250200304, 3.249544054, 3.249771275, 3.250363385, 3.250241072, 3.249718281, 3.249758026, 3.250212024, 3.250233810, 3.249847148, 3.249780306, 3.250104084, 3.250201378, 3.249935698, 3.249819153, 3.250032847, 3.250159243, 3.249991581, 3.249862258, 3.249990198, 3.250116980, 3.250022930, 3.249902425, 3.249968254, 3.250079798, 3.250037147, 3.249936104, 3.249960217, 3.250049902, 3.250040336, 3.249962161, 3.249960703, 3.250027601, 3.250037152, 3.249980909, 3.249965754, 3.250012139, 3.250030899, 3.249993406, 3.249972667, 3.250002271, 3.250023742, 3.250000990, 3.249979748, 3.249996640, 3.250016966, 3.250004984, 3.249986057, 3.249993994, 3.250011223, 3.250006544, 3.249991179, 3.249993291, 3.250006740, 3.250006593, 3.249995032, 3.249993726, 3.250003487, 3.250005816, 3.249997728, 3.249994727, 3.250001297, 3.250004686, 3.249999468, 3.249995912, 3.249999950, 3.250003502, 3.250000476, 3.249997053, 3.249999226, 3.250002433, 3.250000966, 3.249998031, 3.249998927, 3.250001557, 3.250001113, 3.249998802, 3.249998896, 3.250000892, 3.250001058, 3.249999366]

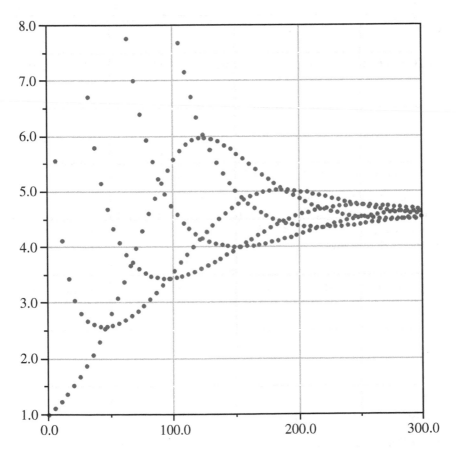

FIGURE 4.10: A graph of $\{b_m\}$, with $k = 1$ and $\alpha = \left(\frac{1}{10}, 0, 0, 0, \frac{9}{10}\right)$.

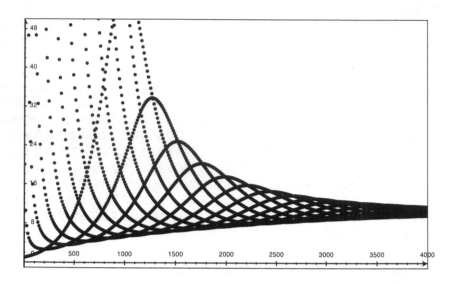

FIGURE 4.11: A graph of $\{b_m\}$, with $\alpha = (0.05, 0, 0, 0, 0, 0, 0, 0, 0, 0, 0.95)$ and $k = 1$.

4.2 A bound for the constant $k_\infty(T)$

We have already seen that, for any α-contraction T, we have $k_0(T) < 1$, and, consequently, $k_\infty(T) = 0$. In particular, it is satisfied for the mapping T as in example 4.3, with $k < 1$. It means that, for $\{b_m\}_{m=0}^{\infty}$ as in theorem 4.3, with $k < 1$, we have

$$\lim_{m \to \infty} b_m = 0.$$

Using (3.10), it is also easy to observe that, for any $k > 1$, we have

$$\lim_{m \to \infty} b_m = \infty.$$

In case of α-nonexpansive mapping T, with $\alpha = (\alpha_1, \alpha_2)$, we proved that

$$k(T^n) \le \frac{1 + \alpha_2}{1 + (-1)^n \alpha_2^{n+1}}.$$

Consequently, we immediately obtain the following:

Corollary 4.2 *If $T : M \to M$ is α-nonexpansive, with $\alpha = (\alpha_1, \alpha_2)$, then*

$$k_\infty(T) \le 1 + \alpha_2. \tag{4.5}$$

Proof. Since $\alpha_2 \in (0, 1)$, we get

$$\lim_{n \to \infty} (-1)^n \alpha_2^{n+1} = 0.$$

Hence,

$$k_\infty(T) = \limsup_{n \to \infty} k(T^n) \leq \lim_{n \to \infty} \frac{1 + \alpha_2}{1 + (-1)^n \alpha_2^{n+1}} = 1 + \alpha_2.$$

$\square$

Example 4.4 *Let $T : \ell_1 \to \ell_1$ be as in example 4.1; that is,*

$$Tx = \left(\frac{1 - (-\alpha_2)^1}{1 - (-\alpha_2)^2} x_2, \frac{1 - (-\alpha_2)^2}{1 - (-\alpha_2)^3} x_3, \dots, \frac{1 - (-\alpha_2)^j}{1 - (-\alpha_2)^{j+1}} x_{j+1}, \dots \right).$$

We proved that T is (α_1, α_2)-nonexpansive and

$$k(T^n) = b_n = \frac{1 + \alpha_2}{1 + (-1)^n \alpha_2^{n+1}},$$

where $\{b_n\}_{n \geq 0}$ is defined as in theorem 4.2, with $k = 1$. Hence,

$$k_\infty(T) = \lim_{n \to \infty} b_n = 1 + \alpha_2.$$

This shows that the evaluation for $k_\infty(T)$ is sharp.

It can be also observed that the sequence of Lipschitz constants $k(T^n) = b_n$ oscillates around the limit $k_\infty(T) = 1 + \alpha_2$ in the following way:

$$k(T^{2n}) = b_{2n} = \frac{1 + \alpha_2}{1 + \alpha_2^{2n+1}} < 1 + \alpha_2$$

and

$$k(T^{2n+1}) = b_{2n+1} = \frac{1 + \alpha_2}{1 - \alpha_2^{2n+2}} > 1 + \alpha_2;$$

see also figures 4.1, 4.2, and 4.3.

Let us also observe that, for any α-nonexpansive mapping T, with $\alpha = (\alpha_1, \alpha_2)$, we have

$$\rho(T^n x, T^n y) \leq \frac{1}{\alpha_1} \rho(x, y).$$

Hence, for α_1 close to 1, the right-hand side of the above inequality is close to 1, whereas for α_1 close to 0 it tends to infinity. Nevertheless, for any α, $k_\infty(T)$ is always less than 2.

In general case of multi-index $\alpha = (\alpha_1, \dots, \alpha_n)$ of length n, we have the following bound, which follows from the theorem 4.3:

Corollary 4.3 *Let (M, ρ) be a metric space. Let $T : M \to M$ be an α-nonexpansive mapping with $\alpha = (\alpha_1, \ldots, \alpha_n)$. Then, $k(T^m) \le b_m$, where the sequence $\{b_m\}_{m=0}^{\infty}$ is defined as follows:*

$$b_m = \begin{cases} 1 & \text{for } m = 0, \\ \dfrac{1}{\sum\limits_{j=1}^{m} \alpha_j b_{m-j}^{-1}} & \text{for } m = 1, \ldots, n, \\ \dfrac{1}{\sum\limits_{i=1}^{n} \alpha_i b_{m-i}^{-1}} & \text{for } m = n+1, n+2, \ldots. \end{cases}$$

As observed in figures 4.6–4.11, the behavior of the sequence $\{b_m\}_{m=0}^{\infty}$ is much more complicated for the multi-index α of length $n \ge 3$ than that for $n = 2$, especially when α has many zeros inside. Moreover, even if we assume that the sequence $\{b_m\}_{m=0}^{\infty}$ converges, the above relation does not give us any information about its limit. Indeed, if g denotes the limit of $\{b_m\}_{m=0}^{\infty}$, then we only get

$$g = \frac{1}{\sum\limits_{i=1}^{n} \alpha_i g^{-1}}.$$

In spite of this, the above-mentioned figures as well as vectors may suggest that the sequence $\{b_m\}_{m=0}^{\infty}$ converges, and it can be also **guessed** that the candidate for the limit is the number

$$g = 1 + (\alpha_2 + \cdots + \alpha_n) + (\alpha_3 + \cdots + \alpha_n) + \cdots + \alpha_n.$$

Indeed,

- for $\alpha = (\frac{1}{3}, \frac{1}{3}, \frac{1}{3})$, see figure 4.6, we get

$$1 + \left(\frac{1}{3} + \frac{1}{3}\right) + \frac{1}{3} = 2;$$

- for $\alpha = (\frac{1}{3}, 0, \frac{2}{3})$, see figure 4.7, we get

$$1 + \left(0 + \frac{2}{3}\right) + \frac{2}{3} = \frac{7}{3} = 2.(3);$$

- for $\alpha = (\frac{1}{4}, \frac{1}{4}, \frac{1}{4}, \frac{1}{4})$, see figure 4.8, we get

$$1 + \left(\frac{1}{4} + \frac{1}{4} + \frac{1}{4}\right) + \left(\frac{1}{4} + \frac{1}{4}\right) + \frac{1}{4} = 2.5;$$

- for $\alpha = \left(\frac{1}{4}, 0, 0, \frac{3}{4}\right)$, see figure 4.9, we get

$$1 + \left(0 + 0 + \frac{3}{4}\right) + \left(0 + \frac{3}{4}\right) + \frac{3}{4} = 3.25.$$

Hence, all the "limit numbers" fit to our samples!

Next observation follows from the examination of our mapping T defined as in example 4.3 (see [65]):

Lemma 4.3 *Let $X = \ell_1$ and T as in example 4.3, with $k = 1$. Then, T is an isometry on the subspace*

$$S := \overline{\text{span}\{e_i : i \geq 2\}},$$

with respect to the new norm $\|\cdot\|_T$ defined for any $x \in \ell_1$ by

$$\|x\|_T = \|x\| + (\alpha_2 + \cdots + \alpha_n)\|Tx\| + \cdots + \alpha_n\|T^{n-1}x\|;$$

that is, for any $x \in S$, we have $\|Tx\|_T = \|x\|_T$.

Proof. Since T is linear and bounded, we conclude that $\|\cdot\|_T$ is a norm that is equivalent to the standard norm $\|\cdot\|$ in ℓ_1. By the equality (4.4), with $k = 1$, for every $x \in S$, $\sum\limits_{i=1}^{n} \alpha_i\|T^i x\| = \|x\|$, and

$$\|Tx\|_T = \sum_{j=1}^{n} \left(\sum_{i=j}^{n} \alpha_i \right) \|T^j x\| = \left(\sum_{i=1}^{n} \alpha_i\|T^i x\| \right) + \sum_{j=2}^{n} \left(\sum_{i=j}^{n} \alpha_i \right) \|T^{j-1} x\|$$

$$= \|x\| + \sum_{j=2}^{n} \left(\sum_{i=j}^{n} \alpha_i \right) \|T^{j-1} x\| = \|x\|_T.$$

$\square$

In view of lemma 4.3, for any $j \geq 0$, we have

$$\|e_{n+j+1}\|_T = \|Te_{n+j+1}\|_T = \cdots = \|T^{n+j}e_{n+j+1}\|_T = \left\| \left(\frac{b_{n+j}}{b_0}, 0, 0, \ldots \right) \right\|_T$$

$$= b_{n+j}.$$

However, by definition of $\|\cdot\|_T$, we get

$$\|e_{n+j+1}\|_T = 1 + (\alpha_2 + \cdots + \alpha_n)\frac{b_{n+j}}{b_{n+j-1}} + (\alpha_3 + \cdots + \alpha_n)\frac{b_{n+j}}{b_{n+j-2}} + \cdots + \alpha_n\frac{b_{n+j}}{b_{j+1}}.$$

Hence,

$$1 + (\alpha_2 + \cdots + \alpha_n)\frac{b_{n+j}}{b_{n+j-1}} + (\alpha_3 + \cdots + \alpha_n)\frac{b_{n+j}}{b_{n+j-2}} + \cdots + \alpha_n\frac{b_{n+j}}{b_{j+1}} = b_{n+j}.$$

Dividing both sides of the above equation by b_{n+j} and defining the sequence

$$d_m = \frac{1}{b_m},$$

we conclude that, for each $j \geq 0$,

$$d_{n+j} + (\alpha_2 + \cdots + \alpha_n)d_{n+j-1} + (\alpha_3 + \cdots + \alpha_n)d_{n+j-2} + \cdots + \alpha_n d_{j+1} = 1.$$

Let us observe that, for

$$d = \frac{1}{g} = \frac{1}{\sum_{j=1}^{n} \left(\sum_{i=j}^{n} \alpha_i \right)},$$

we also have

$$d + (\alpha_2 + \cdots + \alpha_n)d + (\alpha_3 + \cdots + \alpha_n)d + \cdots + \alpha_n d = 1.$$

Substracting both expressions and defining $c_m = d_m - d$, we get

$$c_{n+j} + (\alpha_2 + \cdots + \alpha_n)c_{n+j-1} + (\alpha_3 + \cdots + \alpha_n)c_{n+j-2} + \cdots + \alpha_n c_{j+1} = 0,$$

or, equivalently,

$$c_{n+j} = -(\alpha_2 + \cdots + \alpha_n)c_{n+j-1} - (\alpha_3 + \cdots + \alpha_n)c_{n+j-2} - \cdots - \alpha_n c_{j+1}.$$

Now, our claim is to prove that $\lim_{m \to \infty} c_m = 0$. It is obvious for $n = 1$. Suppose $n \geq 2$. Once we have $c_1, c_2, \ldots, c_{n-1}$, we can define c_m for $m \geq n$ via the relation

$$A \begin{bmatrix} c_{m-1} \\ c_{m-2} \\ c_{m-3} \\ \cdots \\ c_{m-n+1} \end{bmatrix} = \begin{bmatrix} c_m \\ c_{m-1} \\ c_{m-2} \\ \cdots \\ c_{m-n+2} \end{bmatrix},$$

where

$$A = \begin{bmatrix} -(\alpha_2 + \cdots + \alpha_n) & -(\alpha_3 + \cdots + \alpha_n) & -(\alpha_4 + \cdots + \alpha_n) & \cdots & -\alpha_n \\ 1 & 0 & 0 & \cdots & 0 \\ 0 & 1 & 0 & \cdots & 0 \\ \cdots\cdots\cdots\cdots\cdots\cdots\cdots\cdots\cdots\cdots\cdots\cdots\cdots\cdots\cdots\cdots\cdots\cdots\cdots\cdots\cdots\cdots\cdots\cdots & & & & \\ 0 & 0 & & \cdots & 1 & 0 \end{bmatrix}.$$

To show that $\{c_m\}$ tends to zero, it is enough to prove that an equivalent metric for which A is a contraction exists. Then, in view of the classical Banach's Contraction Mapping Principle, the claim will be proved. For this, we need to show that

$$\mathrm{spr}\,(A) = \lim_{m \to \infty} \|A^m\|^{\frac{1}{m}} < 1,$$

what exactly means that all eigenvalues of A have modules strictly less than 1. The characteristic polynomial of A is

$$r(t) = (-1)^{n-1}(t^{n-1} + (\alpha_2 + \cdots + \alpha_n)t^{n-2} + \cdots + \alpha_n).$$

Hence, the eigenvalues of A are exactly the roots of the polynomial

$$p(t) = t^{n-1} + (\alpha_2 + \cdots + \alpha_n)t^{n-2} + \cdots + \alpha_n. \tag{4.6}$$

We shall need several technical lemmas concerning geometry of zeros of polynomials.

Lemma 4.4 *Let $g(t) = b_n t^n + \cdots + b_1 t + b_0$.*

(1) *If $0 < b_n \le b_{n-1} \le \cdots \le b_1 \le b_0$, then every root w of g satisfies $|w| \ge 1$.*

(2) *If $b_n \ge b_{n-1} \ge \cdots \ge b_1 \ge b_0 > 0$, then every root w of g satisfies $|w| \le 1$.*

Proof. (See [5, pp. 180–181]). To prove (1), suppose $|w| < 1$. Then, using triangle inequality, we get

$$
\begin{aligned}
|(1-w)g(w)| &\ge b_0 - |(b_1 - b_0)w| - \cdots - |(b_n - b_{n-1})w^n| - |b_n w^{n+1}| \\
&> b_0 - |b_1 - b_0| - \cdots - |b_n - b_{n-1}| - |b_n| \\
&= b_0 + (b_1 - b_0) + \cdots + (b_n - b_{n-1}) - b_n = 0.
\end{aligned}
$$

Consequently, every root w of g satisfies $|w| \ge 1$.

To prove (2), observe that w is a root of g if and only if $\frac{1}{w}$ is a zero of

$$b_0 t^n + b_1 t^{n-1} + \cdots + b_{n-1} t + b_n.$$

$\square$

Lemma 4.5 *Let $f(t) = a_n t^n + a_{n-1} t^{n-1} + \cdots + a_1 t + a_0$, where $a_n \ge a_{n-1} \ge \cdots \ge a_1 \ge a_0 > 0$. Define the number*

$$v := \max\left\{ \frac{a_{n-1}}{a_n}, \frac{a_{n-2}}{a_{n-1}}, \ldots, \frac{a_1}{a_2}, \frac{a_0}{a_1} \right\}.$$

Then, each root w of polynomial f satisfies $|w| \le v$.

Proof. (See [5, p. 326]). Let $g(t) = a_n v^n t^n + \cdots + a_1 v t + a_0$. If w is a zero of f, then it is also a root of polynomial $g(\frac{t}{v})$; that is, $g(\frac{w}{v}) = 0$. Also, observe that by definition of v, $a_n v^n \ge a_{n-1} v^{n-1} \ge \cdots \ge a_1 v \ge a_0 > 0$. Thus, in view of previous lemma, $\frac{|w|}{v} \le 1$.

$\square$

Actually, in [5, p. 326], the conclusion of lemma 4.5 states that $|w| < v$. However, this is in general false. Consider the polynomial $f(t) = t + 1$ or $f(t) = t^3 + t^2 + t + 1$. Nevertheless, from lemma 4.5, we obtain the following

Corollary 4.4 *Let $f(t) = a_n t^n + a_{n-1} t^{n-1} + \cdots + a_1 t + a_0$, where $a_n > a_{n-1} > \cdots > a_1 > a_0 > 0$. Then each root w of polynomial f satisfies $|w| < 1$.*

Observe that if all α_i are strictly positive, then coefficients of our polynomial p given by formula (4.6) satisfy the assumption of corollary 4.4; that is,

$$1 = \alpha_1 + \alpha_2 + \cdots + \alpha_n > \alpha_2 + \cdots + \alpha_n > \cdots > \alpha_n > 0.$$

Thus, all roots of p are strictly inside the unit complex disk and this implies that our "limit" g is actually the limit of the sequence $\{b_m\}_{m=0}^{\infty}$. Nevertheless, it is only a partial solution of our problem. **What if $\alpha_i \geq 0$ for $i = 2, \ldots, n-1$, $n \geq 3$? Does $\{b_m\}_{m=0}^{\infty}$ converge for any multi-index $\alpha = (\alpha_1, \ldots, \alpha_n)$? If the answer is affirmative, is g the limit of $\{b_m\}_{m=0}^{\infty}$?** Our samples strongly suggest that the answer is **YES!** To prove it, we shall need one more result concerning roots of polynomial. Maybe the following lemma is known, but we were not able to find any references about it.

Lemma 4.6 (Pérez García and Piasecki [65]) *Let $n \geq 2$ and $1 = a_n > a_{n-1} \geq a_{n-2} \geq \cdots \geq a_1 \geq a_0 > 0$. If z is any root of the polynomial*

$$f(t) = a_n t^n + a_{n-1} t^{n-1} + \cdots + a_1 t + a_0, \tag{4.7}$$

then $|z| < 1$.

Proof. Let us define

$$v = \max\left\{\frac{a_{n-1}}{a_n}, \frac{a_{n-2}}{a_{n-1}}, \ldots, \frac{a_1}{a_2}, \frac{a_0}{a_1}\right\}.$$

To prove our claim, we will construct a polynomial $q(t) = r(t)f(t)$ for which assumption of corollary 4.4 holds. Then, all roots of q, which contain roots of f, will be contained strictly inside the unit complex disk.

Suppose that in our polynomial (4.7) we have $a_j = a_{j+1}$ for some j. We can select c such that $0 < c < 1 - a_{n-1} < 1$ and observe that the polynomial

$$q_1(t) = (t + c)f(t) = a_n t^{n+1} + (a_{n-1} + ca_n)t^n + \cdots + (a_0 + ca_1)t + ca_0$$

has the following properties:

1. The degree of q_1 equals the degree of f plus 1.

2. All coefficients of q_1 are positive and nonincreasing. Indeed, $1 = a_n > a_{n-1} + ca_n = a_{n-1} + c$, and for $j = 1, \ldots, n-1$, we have $a_j + ca_{j+1} \geq a_{j-1} + ca_j$. Also, $a_0 + ca_1 > ca_0 > 0$.

3. Suppose that for some k we have $a_{j-1} < a_j = a_{j+1} = \cdots = a_{j+k-1} = a_{j+k} < a_{j+k+1}$ in our polynomial f. Then, we have the following inequalities for the coefficients of q_1:

$$a_{j-1} + ca_j < a_j + ca_{j+1} = a_{j+1} + ca_{j+2} = \cdots = a_{j+k-1} + ca_{j+k}$$
$$< a_{j+k} + ca_{j+k+1}.$$

If the polynomial q_1 still has some equal coefficients, then we apply the same procedure to q_1 to generate a polynomial q_2. After at most $n-1$ steps, we will obtain the desired polynomial $q(t) = r(t)f(t)$ such that all its coefficients are strictly decreasing.

□

The coefficients of our polynomial p are positive and nonincreasing. Moreover, $1 = \alpha_1 + \alpha_2 + \cdots + \alpha_n > \alpha_2 + \cdots + \alpha_n$ because $\alpha_1 > 0$. Finally, applying lemma 4.6 to the polynomial p, we finish the proof.

Reassuming, all the above lead us to the following

Theorem 4.4 (Pérez García and Piasecki [65]) *Let $\alpha = (\alpha_1, \ldots, \alpha_n)$ be a multi-index. If $\{b_m\}_{m=0}^{\infty}$ is defined as*

$$
b_m = \begin{cases}
1 & \text{for} \quad m = 0, \\[2mm]
\dfrac{1}{\sum\limits_{j=1}^{m} \alpha_j b_{m-j}^{-1}} & \text{for} \quad m = 1, \ldots, n, \\[4mm]
\dfrac{1}{\sum\limits_{i=1}^{n} \alpha_i b_{m-i}^{-1}} & \text{for} \quad m = n+1, n+2, \ldots,
\end{cases}
$$

then

$$
\lim_{m \to \infty} b_m = \sum_{j=1}^{n} \left(\sum_{i=j}^{n} \alpha_i \right) = 1 + \alpha_2 + 2\alpha_3 + 3\alpha_4 + \cdots + (n-1)\alpha_n
$$

$$
= \alpha_1 + 2\alpha_2 + 3\alpha_3 + \cdots + n\alpha_n.
$$

Corollary 4.5 *If $T : M \to M$ is an $(\alpha_1, \ldots, \alpha_n)$-nonexpansive mapping, then*

$$
k_\infty(T) \leq 1 + \alpha_2 + 2\alpha_3 + 3\alpha_4 + \cdots + (n-1)\alpha_n
$$
$$
= \alpha_1 + 2\alpha_2 + 3\alpha_3 + \cdots + n\alpha_n.
$$

Proof. It is a consequence of corollary 4.3 and theorem 4.4.

□

To see that the estimate for $k_\infty(T)$ is sharp, it is enough to consider the mapping T from example 4.3 with $k = 1$. Also, observe that for any $n \geq 2$ and any $(\alpha_1, \ldots, \alpha_n)$-nonexpansive mapping T, $k_\infty(T)$ is always less than the length of multi-index α; that is, $k_\infty(T) < n$.

4.3 Moving averages in Banach spaces

The study of mean nonexpansive mappings leads us to the moving average problem.

Let us recall that for any α-nonexpansive mapping $T : M \to M$ with $\alpha = (\alpha_1, \ldots, \alpha_n)$ we proved that $k(T^m) \leq b_m$, where the sequence $\{b_m\}_{m=0}^\infty$ is defined as follows:

$$
b_m = \begin{cases}
1 & \text{for} \quad m = 0, \\[2mm]
\dfrac{1}{\sum\limits_{j=1}^{m} \alpha_j b_{m-j}^{-1}} & \text{for} \quad m = 1, \ldots, n, \\[6mm]
\dfrac{1}{\sum\limits_{i=1}^{n} \alpha_i b_{m-i}^{-1}} & \text{for} \quad m = n+1, n+2, \ldots.
\end{cases}
$$

We also showed that this bound is sharp. We used the mapping $T : \ell_1 \to \ell_1$ defined for $x = (x_1, x_2, \ldots) \in \ell_1$ by

$$
Tx = \left(\frac{b_1}{b_0} x_2, \frac{b_2}{b_1} x_3, \ldots, \frac{b_j}{b_{j-1}} x_{j+1}, \ldots \right).
$$

Later, we studied the asymptotic behavior of the sequence of Lipschitz constants for mean nonexpansive mappings in term of the constant $k_\infty(T)$. We defined $d_m = 1/b_m$ to obtain a relation:

$$
d_m = \begin{cases}
1 & \text{for} \quad m = 0, \\[2mm]
\sum\limits_{j=1}^{m} \alpha_j d_{m-j} & \text{for} \quad m = 1, \ldots, n, \\[6mm]
\sum\limits_{i=1}^{n} \alpha_i d_{m-i} & \text{for} \quad m = n+1, n+2, \ldots.
\end{cases}
$$

We proved one result concerning localization of roots of polynomials (lemma 4.6). We used this to show that the eigenvalues of a special matrix lie strictly inside the complex unit disk. Using also a special property of the mapping T, we finally obtained

$$
\lim_{m \to \infty} d_m = \frac{1}{\sum_{j=1}^{n} \left(\sum_{i=j}^{n} \alpha_i \right)},
$$

and, consequently,

$$
\lim_{m \to \infty} b_m = \sum_{j=1}^{n} \left(\sum_{i=j}^{n} \alpha_i \right).
$$

If we follow the tricks used in the proof of theorem 4.4, this result can be easily extended to the *moving average problem*: find the limit of the sequence $\{d_m\}_{m=0}^\infty$, where $d_0, \ldots, d_{n-1}$ are arbitrary numbers and for $m \geq n$

$$
d_m = \alpha_1 d_{m-1} + \cdots + \alpha_n d_{m-n}.
$$

Theorem 4.5 (Pérez García and Piasecki) *Let* $\alpha = (\alpha_1, \ldots, \alpha_n)$ *be a multi-index. Let* $d_0, \ldots, d_{n-1}$ *be arbitrary numbers and* d_m *be defined for* $m \geq n$ *by*

$$
d_m = \sum_{i=1}^{n} \alpha_i d_{m-i}.
$$

Then,

$$\lim_{m \to \infty} d_m = \frac{1}{\sum_{i=1}^{n} \sum_{j=i}^{n} \alpha_j} \sum_{i=1}^{n} \left(\sum_{j=i}^{n} \alpha_j \right) d_{n-i}.$$

Proof. It is enough to note that the following relation holds under our assumption of the sequence $\{d_m\}_{m=0}^{\infty}$:

$$(\alpha_1 + \cdots + \alpha_n)d_{n+j} + (\alpha_2 + \cdots + \alpha_n)d_{n+j-1} + \cdots + \alpha_n d_{j+1}$$
$$= (\alpha_1 + \cdots + \alpha_n)d + (\alpha_2 + \cdots + \alpha_n)d + \cdots + \alpha_n d,$$

where

$$d = \frac{1}{\sum_{i=1}^{n} \sum_{j=i}^{n} \alpha_j} \sum_{i=1}^{n} \left(\sum_{j=i}^{n} \alpha_j \right) d_{n-i}.$$

Consequently,

$$(d_{n+j} - d) + (\alpha_2 + \cdots + \alpha_n)(d_{n+j-1} - d) + \cdots + \alpha_n(d_{j+1} - d) = 0.$$

If we set $c_m = d_m - d$, we get that $\lim_{m \to \infty} c_m = 0$ as in the proof of the theorem 4.4.

$\square$

Let us pass now to the general case of Banach spaces. Suppose X is a Banach space. Fix a multi-index $\alpha = (\alpha_1, \ldots, \alpha_n)$. Let $x_0, \ldots, x_{n-1}$ be arbitrary elements of X and $\{x_m\}$ be defined for $m \geq n$ as

$$x_m = \alpha_1 x_{m-1} + \alpha_2 x_{m-2} + \cdots + \alpha_n x_{m-n}.$$

Now, if we put $y_m = x_m - x$, with

$$x = \frac{1}{\sum_{i=1}^{n} \sum_{j=i}^{n} \alpha_j} \sum_{i=1}^{n} \left(\sum_{j=i}^{n} \alpha_j \right) x_{n-i},$$

we see that $\{y_m\}_{m=0}^{\infty}$ converges weakly to 0. Since $\{y_m\}_{m=0}^{\infty} \subset$ span $\{x_0, \ldots, x_{n-1}\}$, we have that the convergence is strong. Summarizing the above observations, we get the following:

Theorem 4.6 (Pérez García and Piasecki) *Let* $\alpha = (\alpha_1, \ldots, \alpha_n)$ *be as above. Let* $x_0, \ldots, x_{n-1}$ *be arbitrary elements of a Banach space* X *and* x_m *be defined for* $m \geq n$ *by*

$$x_m = \sum_{i=1}^{n} \alpha_i x_{m-i}.$$

Then,

$$\lim_{m \to \infty} x_m = \frac{1}{\sum_{i=1}^{n} \sum_{j=i}^{n} \alpha_j} \sum_{i=1}^{n} \left(\sum_{j=i}^{n} \alpha_j \right) x_{n-i}.$$

4.4   A bound for the constant $k_0(T)$

Let us recall that, for any α-lipschitzian mapping $T : M \to M$ with $\alpha = (\alpha_1, \alpha_2)$ and $k > 0$, we have the following bound for the sequence of consecutive Lipschitz constants of the iterates of T:

$$k(T^n) \leq b_n,$$

where

$$b_n = \begin{cases} 1 & \text{for} \quad n = 0, \\ \frac{k}{\alpha_1} & \text{for} \quad n = 1, \\ \frac{k}{\alpha_1 b_{n-1}^{-1} + \alpha_2 b_{n-2}^{-1}} & \text{for} \quad n = 2, 3, \ldots. \end{cases}$$

We have also proved that the following relation for $\{b_n\}_{n=0}^{\infty}$ holds:

$$b_n = k^n \frac{2^{n+1}\sqrt{\Delta}}{\left(\alpha_1 + \sqrt{\Delta}\right)^{n+1} - \left(\alpha_1 - \sqrt{\Delta}\right)^{n+1}},$$

where $\Delta = \alpha_1^2 + 4\alpha_2 k$.

Now we use the above facts to prove the following:

Theorem 4.7 *Let $T : M \to M$ be (α_1, α_2)-lipschitzian mapping for the constant $k > 0$. Then,*

$$k_0(T) \leq k \frac{2}{\alpha_1 + \sqrt{\Delta}},$$

with Δ as above.

Proof. One can notice that, for each $n \geq 1$,

$$a_n \leq (b_n)^{\frac{1}{n}} \leq c_n,$$

where $\{b_n\}_{n=0}^{\infty}$ is given as above,

$$a_n = k \left(\frac{2^{n+1}\sqrt{\Delta}}{2 \left(\alpha_1 + \sqrt{\Delta}\right)^{n+1}} \right)^{\frac{1}{n}} = k \cdot \frac{2}{\alpha_1 + \sqrt{\Delta}} \cdot \left(\frac{\sqrt{\Delta}}{\alpha_1 + \sqrt{\Delta}} \right)^{\frac{1}{n}}$$

and

$$c_n = \begin{cases} k \cdot \left(\frac{2^{n+1}\sqrt{\Delta}}{(\alpha_1 + \sqrt{\Delta})^{n+1}} \right)^{\frac{1}{n}} & \text{for } n = 2l, \ l = 0, 1, \ldots \\ k \cdot \left(\frac{2^{n+1}\sqrt{\Delta}}{2\alpha_1 (\alpha_1 + \sqrt{\Delta})^n} \right)^{\frac{1}{n}} & \text{for } n = 2l + 1, \ l = 0, 1, \ldots \end{cases}$$

$$= \begin{cases} k \cdot \frac{2}{\alpha_1 + \sqrt{\Delta}} \cdot \left(\frac{2\sqrt{\Delta}}{\alpha_1 + \sqrt{\Delta}} \right)^{\frac{1}{n}} & \text{for } n = 2l, \ l = 0, 1, \ldots \\ k \cdot \frac{2}{\alpha_1 + \sqrt{\Delta}} \cdot \left(\frac{\sqrt{\Delta}}{\alpha_1} \right)^{\frac{1}{n}} & \text{for } n = 2l + 1, \ l = 0, 1, \ldots. \end{cases}$$

Obviously,

$$\lim_{n\to\infty} a_n = \lim_{n\to\infty} c_n = k\frac{2}{\alpha_1 + \sqrt{\Delta}}.$$

Hence,

$$\lim_{n\to\infty} (b_n)^{\frac{1}{n}} = k\frac{2}{\alpha_1 + \sqrt{\Delta}}.$$

$\square$

Example 4.5 *Let* $T : \ell_1 \to \ell_1$ *be defined as in example 4.2; that is,*

$$T(x_1, x_2, \dots) = \left(\frac{b_1}{b_0}x_2, \frac{b_2}{b_1}x_3, \dots, \frac{b_j}{b_{j-1}}x_{j+1}, \dots\right).$$

We proved that T *is* (α_1, α_2)*-lipschitzian for the constant* $k > 0$*, and for* $n = 1, 2, \dots$*, we have*

$$k(T^n) = b_n.$$

In particular, the above implies that

$$k_0(T) = k\frac{2}{\alpha_1 + \sqrt{\Delta}}. \tag{4.8}$$

Let us observe that, in this special case, we can give an explicit formula for an equivalent norm in ℓ_1*, denoted by* $\|\cdot\|_T$*, with respect to which* T *satisfies the Lipschitz condition with the constant* $2k/\left(\alpha_1 + \sqrt{\Delta}\right)$*; for* $x = (x_1, x_2, \dots) \in \ell_1$*, we put*

$$\|x\|_T = \sum_{i=1}^{\infty} \frac{2\sqrt{\Delta}\left(\alpha_1 + \sqrt{\Delta}\right)^{i-1}}{\left(\alpha_1 + \sqrt{\Delta}\right)^i - \left(\alpha_1 - \sqrt{\Delta}\right)^i} |x_i|.$$

It is easy to observe that, for $i \geq 1$*,*

$$\frac{2\sqrt{\Delta}\left(\alpha_1 + \sqrt{\Delta}\right)^{i-1}}{\left(\alpha_1 + \sqrt{\Delta}\right)^i - \left(\alpha_1 - \sqrt{\Delta}\right)^i} = \frac{1}{\alpha_1 + \sqrt{\Delta}} \cdot \frac{2\sqrt{\Delta}}{1 - \left(\frac{\alpha_1 - \sqrt{\Delta}}{\alpha_1 + \sqrt{\Delta}}\right)^i}$$

$$\geq \frac{1}{\alpha_1 + \sqrt{\Delta}} \cdot \frac{2\sqrt{\Delta}}{1 - \frac{\alpha_1 - \sqrt{\Delta}}{\alpha_1 + \sqrt{\Delta}}} = 1,$$

and

$$\frac{2\sqrt{\Delta}\left(\alpha_1 + \sqrt{\Delta}\right)^{i-1}}{\left(\alpha_1 + \sqrt{\Delta}\right)^i - \left(\alpha_1 - \sqrt{\Delta}\right)^i} = \frac{1}{\alpha_1 + \sqrt{\Delta}} \cdot \frac{2\sqrt{\Delta}}{1 - \left(\frac{\alpha_1 - \sqrt{\Delta}}{\alpha_1 + \sqrt{\Delta}}\right)^i}$$

$$\leq \frac{1}{\alpha_1 + \sqrt{\Delta}} \cdot \frac{2\sqrt{\Delta}}{1 - \left(\frac{\alpha_1 - \sqrt{\Delta}}{\alpha_1 + \sqrt{\Delta}}\right)^2}$$

$$= \frac{\alpha_1 + \sqrt{\Delta}}{2\alpha_1}.$$

Hence, $\|\cdot\|_T$ is equivalent to $\|\cdot\|$, and for any $x \in \ell_1$,

$$\|x\| \leq \|x\|_T \leq \frac{\alpha_1 + \sqrt{\Delta}}{2\alpha_1} \|x\|.$$

Further, for every $x \in \ell_1$, we have

$$
\begin{aligned}
\|Tx\|_T &= \sum_{i=1}^{\infty} 2k \frac{2\sqrt{\Delta}\left(\alpha_1 + \sqrt{\Delta}\right)^{i-1}}{\left(\alpha_1 + \sqrt{\Delta}\right)^{i+1} - \left(\alpha_1 - \sqrt{\Delta}\right)^{i+1}} |x_{i+1}| \\
&= \frac{2k}{\alpha_1 + \sqrt{\Delta}} \sum_{i=1}^{\infty} \frac{2\sqrt{\Delta}\left(\alpha_1 + \sqrt{\Delta}\right)^{i}}{\left(\alpha_1 + \sqrt{\Delta}\right)^{i+1} - \left(\alpha_1 - \sqrt{\Delta}\right)^{i+1}} |x_{i+1}| \\
&\leq \frac{2k}{\alpha_1 + \sqrt{\Delta}} \sum_{i=1}^{\infty} \frac{2\sqrt{\Delta}\left(\alpha_1 + \sqrt{\Delta}\right)^{i-1}}{\left(\alpha_1 + \sqrt{\Delta}\right)^{i} - \left(\alpha_1 - \sqrt{\Delta}\right)^{i}} |x_i| \\
&= \frac{2k}{\alpha_1 + \sqrt{\Delta}} \|x\|_T,
\end{aligned}
$$

and by (1.10),

$$k_{\|\cdot\|_T}(T) = k \frac{2}{\alpha_1 + \sqrt{\Delta}}.$$

Now let $\alpha = (\alpha_1, \ldots, \alpha_n)$ be a multi-index, $k > 0$ and $\{b_m\}_{m=0}^{\infty}$ as in theorem 4.3; that is,

$$
b_m = \begin{cases}
1 & \text{for} \quad m = 0, \\
\dfrac{k}{\sum\limits_{j=1}^{m} \alpha_j b_{m-j}^{-1}} & \text{for} \quad m = 1, \ldots, n, \\
\dfrac{k}{\sum\limits_{i=1}^{n} \alpha_i b_{m-i}^{-1}} & \text{for} \quad m = n+1, n+2, \ldots.
\end{cases}
$$

Suppose that $M = \ell_1$ is the space of absolutely summable sequences with the standard norm $\|\cdot\|$ and $T : \ell_1 \to \ell_1$ is a linear mapping defined for $x = (x_1, x_2, \ldots) \in \ell_1$ by

$$Tx = \left(\frac{b_1}{b_0} x_2, \frac{b_2}{b_1} x_3, \ldots, \frac{b_j}{b_{j-1}} x_{j+1}, \ldots\right).$$

In example 4.3, it was shown that T is α-lipschitzian with a constant k and $k(T^m) = b_m$ for every $m \geq 0$; in other words, if we define

$$\|T^m\| = \sup\left\{\|T^m x\| / \|x\| : x \neq 0\right\},$$

then $\|T^m\| = b_m$. In particular, this implies that $\lim_{m \to \infty} (b_m)^{1/m}$ exists. In the same example, it was also proved that, for every $x = (x_1, x_2, \dots) \in \ell_1$,

$$\sum_{j=1}^{n} \alpha_j \|T^j x\| = k \sum_{j=2}^{\infty} |x_j|.$$

If we define

$$E_m = \overline{\operatorname{span}\{e_i : i \geq m+1\}},$$

the last fact means that, for every $x \subset E_1$,

$$\sum_{j=1}^{n} \alpha_j \|T^j x\| = k \|x\|. \tag{4.9}$$

Lemma 4.7 (Pérez García and Piasecki, [66]) *Let $\alpha = (\alpha_1, \dots, \alpha_n)$ be a multi-index, $k > 0$, $\{b_m\}_{m=0}^{\infty}$ and $T : \ell_1 \to \ell_1$ as above. Let $\|\cdot\|_T$ be a norm on ℓ_1 defined for every $x \in \ell_1$ by*

$$\|x\|_T = \left(\alpha_1 + \alpha_2 g + \cdots + \alpha_n g^{n-1}\right) \|x\|$$
$$+ \left(\alpha_2 + \alpha_3 g + \cdots + \alpha_n g^{n-2}\right) \|Tx\|$$
$$+ \left(\alpha_3 + \alpha_4 g + \cdots + \alpha_n g^{n-3}\right) \|T^2 x\| + \dots$$
$$+ \alpha_n \|T^{n-1} x\|,$$

where g is the unique positive solution of the equation

$$\alpha_1 g + \alpha_2 g^2 + \cdots + \alpha_n g^n = k.$$

Then, $\|\cdot\|_T$ is equivalent to $\|\cdot\|$, and for every $m = 1, 2, \dots$, we have

$$k_{\|\cdot\|_T}(T^m) = g^m.$$

Proof. Since T is a bounded linear operator we conclude that $\|\cdot\|_T$ is a norm equivalent to $\|\cdot\|$.

For any $x \in \ell_1$, we have

$$\|Tx\|_T = \left(\alpha_1 + \alpha_2 g + \cdots + \alpha_n g^{n-1}\right) \|Tx\|$$
$$+ \left(\alpha_2 + \alpha_3 g + \cdots + \alpha_n g^{n-2}\right) \|T^2 x\| + \dots$$
$$+ \alpha_n \|T^n x\|$$
$$= \alpha_1 \|Tx\| + \alpha_2 \|T^2 x\| + \cdots + \alpha_n \|T^n x\|$$
$$+ (\alpha_2 g + \cdots + \alpha_n g^{n-1}) \|Tx\|$$
$$+ (\alpha_3 g + \cdots + \alpha_n g^{n-2}) \|T^2 x\| + \dots$$
$$+ \alpha_n g \|T^{n-1} x\|$$

$$\leq k\|x\|$$
$$+ (\alpha_2 g + \cdots + \alpha_n g^{n-1})\|Tx\|$$
$$+ (\alpha_3 g + \cdots + \alpha_n g^{n-2})\|T^2 x\| + \ldots$$
$$+ \alpha_n g\|T^{n-1} x\|$$
$$= (\alpha_1 g + \alpha_2 g^2 + \cdots + \alpha_n g^n)\|x\|$$
$$+ (\alpha_2 g + \cdots + \alpha_n g^{n-1})\|Tx\|$$
$$+ (\alpha_3 g + \cdots + \alpha_n g^{n-2})\|T^2 x\| + \ldots$$
$$+ \alpha_n g\|T^{n-1} x\|$$
$$= g\|x\|_T.$$

In view of (4.9), for every $x \in E_1$, $\|Tx\|_T = g\|x\|_T$; in general, for every $x \in E_m$, $\|T^m x\|_T = g^m \|x\|_T$, hence for every $m \geq 1$,

$$\|T^m\|_T = g^m,$$

where $\|T^m\|_T = \sup\{\|T^m x\|_T / \|x\|_T : x \neq 0\}$. Consequently,

$$k_{\|\cdot\|_T}(T^m) = g^m,$$

as we desired.

$\square$

Theorem 4.8 (Pérez García and Piasecki, [66]) *For the sequence $\{b_m\}$ as above we have*

$$\lim_{m \to \infty} (b_m)^{1/m} = g,$$

where g is the unique positive solution of the equation

$$\alpha_1 g + \alpha_2 g^2 + \cdots + \alpha_n g^n = k.$$

Proof. Since $\|\cdot\|$ and $\|\cdot\|_T$ are equivalent norms, we obtain

$$\lim_{m \to \infty} (b_m)^{1/m} = \lim_{m \to \infty} \|T^m\|^{1/m} = \lim_{m \to \infty} \|T^m\|_T^{1/m} = g.$$

$\square$

Finally, in view of theorem 4.3, we get following

Corollary 4.6 *Let (M, ρ) be a metric space and $T : M \to M$ an α-lipschitzian mapping with $\alpha = (\alpha_1, \ldots, \alpha_n)$ and $k > 0$. Then,*

$$k_0(T) \leq g,$$

where g is given as above.

In fact, as we can see in the proof of lemma 4.7, this estimation is sharp. Moreover, by following its proof, it can be easily observed that for any α-lipschitzian mapping $T : M \to M$ with $\alpha = (\alpha_1, \ldots, \alpha_n)$ and $k > 0$, we can define a metric d_T by

$$
\begin{aligned}
d_T(x, y) = {} & \left(\alpha_1 + \alpha_2 g + \cdots + \alpha_n g^{n-1}\right) \rho(x, y) \\
& + \left(\alpha_2 + \alpha_3 g + \cdots + \alpha_n g^{n-2}\right) \rho(Tx, Ty) \\
& + \left(\alpha_3 + \alpha_4 g + \cdots + \alpha_n g^{n-3}\right) \rho(T^2 x, T^2 y) + \ldots \\
& + \alpha_n \rho(T^{n-1} x, T^{n-1} y),
\end{aligned}
$$

which is equivalent to the metric ρ and such that, for all $x, y \in M$, we have

$$
d_T(Tx, Ty) \leq g d_T(x, y).
$$

Consequently, $k_{d_T}(T) \leq g$.

Let us observe that only in the case of α-nonexpansive mapping, the above metric coincides with a metric given by formula (3.11).

4.5 More about $k(T^n)$, $k_0(T)$, and $k_\infty(T)$

Consider first the situation on the half-line $[0, \infty)$:

Lemma 4.8 *Let* $\alpha = (\alpha_1, \alpha_2)$, $k \in (0, 1]$, *and* $\{b_n\}_{n=0}^{\infty}$ *be as in theorem 4.2; that is,*

$$
b_n = \begin{cases}
1 & \text{for} \quad n = 0, \\
\frac{k}{\alpha_1} & \text{for} \quad n = 1, \\
\frac{k}{\alpha_1 b_{n-1}^{-1} + \alpha_2 b_{n-2}^{-1}} & \text{for} \quad n = 2, 3, \ldots.
\end{cases}
$$

Let $\{x_n\}_{n=0}^{\infty}$ *be defined as follows,*

$$
x_n = \begin{cases}
\frac{1}{b_0} & \text{for} \quad n = 0, \\
x_{n-1} + \frac{1}{b_n} & \text{for} \quad n = 1, 2, \ldots.
\end{cases}
$$

Then, the function $f : [0, \infty) \to [0, \infty)$ *given by*

$$
f(t) = \begin{cases}
0, & \text{if } t \in [0, x_0], \\
\frac{b_{n+1}}{b_n}(t - x_{n+1}) + x_n, & \text{if } t \in [x_n, x_{n+1}], \quad n \geq 0
\end{cases}
$$

is (α_1, α_2)-*lipschitzian for the constant* k *and* $k(f^n) = b_n$ *for* $n \geq 1$ *(see figure 4.12).*

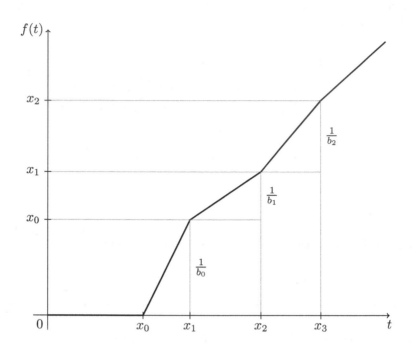

FIGURE 4.12: A graph of function f.

Proof. It is clear that, for $n \geq 0$, we have

$$x_n = \sum_{i=0}^{n} \frac{1}{b_i}.$$

If $k < 1$, then $\lim_{n\to\infty} \sqrt[n]{b_n} < 1$. This implies that $\lim_{n\to\infty} x_n = \infty$. If $k = 1$, then also $\lim_{n\to\infty} x_n = \infty$ because $\lim_{n\to\infty} 1/b_n = 1/(1 + \alpha_2) > 0$. Hence, for $k \in (0, 1]$, the function f is defined on $[0, \infty)$. It is not true for $k > 1$ because $\lim_{n\to\infty} \sqrt[n]{b_n} > 1$, and this implies that $\lim_{n\to\infty} x_n < \infty$.

The function f is well defined because, for each $i \geq 0$ (we set $x_{-1} = 0$), we have

$$f(x_i) = \frac{b_{i+1}}{b_i}(x_i - x_{i+1}) + x_i = \frac{b_{i+1}}{b_i} \cdot \left(-\frac{1}{b_{i+1}}\right) + x_i = -\frac{1}{b_i} + x_i = x_{i-1},$$

and

$$f(x_{i+1}) = \frac{b_{i+1}}{b_i}(x_{i+1} - x_{i+1}) + x_i = x_i.$$

This in particular implies that $f([x_i, x_{i+1}]) = [x_{i-1}, x_i]$ for each $i \geq 0$.

We shall prove that function f is (α_1, α_2)-lipschitzian for the constant k.

Let $t \in [x_n, x_{n+1}]$ and $s \in [x_{n+j}, x_{n+j+1}]$, $s \geq t$, $j \geq 0$, and $n \geq 1$. Hence, there exist $\lambda_1, \lambda_2 \in [0, 1]$ such that $t = x_n + \lambda_1/b_{n+1}$ and $s = x_{n+j} + \lambda_2/b_{n+j+1}$. Then,

$$f(t) = \frac{b_{n+1}}{b_n} \left(x_n + \lambda_1 \frac{1}{b_{n+1}} - x_{n+1} \right) + x_n = (\lambda_1 - 1) \frac{1}{b_n} + x_n,$$

and, by analogy,

$$f(s) = (\lambda_2 - 1) \frac{1}{b_{n+j}} + x_{n+j}.$$

Further,

$$f^2(t) = \frac{b_n}{b_{n-1}} \left(f(t) - x_n \right) + x_{n-1} = (\lambda_1 - 1) \frac{1}{b_{n-1}} + x_{n-1},$$

and

$$f^2(s) = (\lambda_2 - 1) \frac{1}{b_{n+j-1}} + x_{n+j-1}.$$

Thus,

$$\alpha_1 \left(f(s) - f(t) \right) + \alpha_2 \left(f^2(s) - f^2(t) \right)$$

$$= (\lambda_2 - 1) \left(\alpha_1 \frac{1}{b_{n+j}} + \alpha_2 \frac{1}{b_{n+j-1}} \right) + (1 - \lambda_1) \left(\alpha_1 \frac{1}{b_n} + \alpha_2 \frac{1}{b_{n-1}} \right)$$

$$+ \alpha_1 x_{n+j} + \alpha_2 x_{n+j-1} - \alpha_1 x_n - \alpha_2 x_{n-1}.$$

By definitions of $\{x_n\}_{n \geq 0}$ and $\{b_n\}_{n \geq 0}$, we have

$$\alpha_1 x_{n+j} + \alpha_2 x_{n+j-1}$$

$$= \alpha_1 \left(\sum_{i=0}^{n+j} \frac{1}{b_i} \right) + \alpha_2 \left(\sum_{i=0}^{n+j-1} \frac{1}{b_i} \right)$$

$$= \alpha_1 + \sum_{i=1}^{n+j} \left(\alpha_1 \frac{1}{b_i} + \alpha_2 \frac{1}{b_{i-1}} \right)$$

$$= \alpha_1 + \sum_{i=1}^{n+j} k \frac{1}{b_{i+1}}$$

$$= \alpha_1 + k x_{n+j} + k \frac{1}{b_{n+j+1}} - k \frac{1}{b_0} - k \frac{1}{b_1}$$

and

$$-\alpha_1 x_n - \alpha_2 x_{n-1} = -\alpha_1 - k x_n - k \frac{1}{b_{n+1}} + k \frac{1}{b_0} + k \frac{1}{b_1}.$$

Consequently,

$$\alpha_1 \left(f(s) - f(t) \right) + \alpha_2 \left(f^2(s) - f^2(t) \right)$$

$$= k \left(x_{n+j} - x_n + \lambda_2 \frac{1}{b_{n+j+1}} - \lambda_1 \frac{1}{b_{n+1}} \right)$$

$$= k(s - t).$$

If $s, t \in [x_0, x_1]$, $s \geq t$, then $f(s) = \frac{b_1}{b_0}(s - x_1) + x_0$, $f(t) = \frac{b_1}{b_0}(t - x_1) + x_0$, and $f^2(s) = f^2(t) = 0$. Hence,

$$\alpha_1 \left(f(s) - f(t) \right) + \alpha_2 \left(f^2(s) - f^2(t) \right) = \alpha_1 \frac{b_1}{b_0}(s - t) = k(s - t).$$

If $s, t \in [0, x_0]$, then $f(s) = f(t) = f^2(s) = f^2(t) = 0$.
If $t \in [0, x_0]$ and $s \in [x_0, x_1]$, then $f(t) = f^2(t) = f^2(s) = 0$, and we obtain

$$\alpha_1 \left(f(s) - f(t) \right) + \alpha_2 \left(f^2(s) - f^2(t) \right)$$

$$= \alpha_1 \left(\frac{b_1}{b_0}(s - x_1) + x_0 \right)$$

$$= \alpha_1 \left(\frac{k}{\alpha_1} \left(s - x_0 - \frac{\alpha_1}{k} \right) + 1 \right)$$

$$= k(s - x_0)$$

$$\leq k(s - t).$$

If $t \in [0, x_0]$ and $s \in [x_n, x_{n+1}]$, $n \geq 1$, then

$$\alpha_1 \left(f(s) - f(t) \right) + \alpha_2 \left(f^2(s) - f^2(t) \right)$$

$$= \alpha_1 \left(\frac{b_{n+1}}{b_n}(s - x_{n+1}) + x_n \right) + \alpha_2 \left(\frac{b_{n+1}}{b_{n-1}}(s - x_{n+1}) + x_{n-1} \right)$$

$$= (s - x_{n+1}) \left(\alpha_1 \frac{b_{n+1}}{b_n} + \alpha_2 \frac{b_{n+1}}{b_{n-1}} \right) + \alpha_1 x_n + \alpha_2 x_{n-1}$$

$$= k(s - x_{n+1}) + \alpha_1 \sum_{i=0}^{n} \frac{1}{b_i} + \alpha_2 \sum_{i=0}^{n-1} \frac{1}{b_i}$$

$$= k(s - x_{n+1}) + \sum_{i=1}^{n} \left(\alpha_1 \frac{1}{b_i} + \alpha_2 \frac{1}{b_{i-1}} \right) + \alpha_1 \frac{1}{b_0}$$

$$= ks - k \sum_{i=0}^{n+1} \frac{1}{b_i} + k \sum_{i=1}^{n} \frac{1}{b_{i+1}} + \alpha_1$$

$$= ks - k \frac{1}{b_0} - k \frac{1}{b_1} + \alpha_1$$

$$= k(s - 1)$$

$$\leq k(s - t).$$

If $t \in [x_0, x_1]$, $s \in [x_n, x_{n+1}]$, and $n \geq 1$, then $f^2(t) = 0$, and

$$\alpha_1 \left(f(s) - f(t) \right) + \alpha_2 \left(f^2(s) - f^2(t) \right)$$

$$= \alpha_1 \left(\frac{b_{n+1}}{b_n} (s - x_{n+1}) + x_n - \frac{b_1}{b_0} (t - x_1) - x_0 \right)$$

$$+ \alpha_2 \left(\frac{b_{n+1}}{b_{n-1}} (s - x_{n+1}) + x_{n-1} \right)$$

$$= (s - x_{n+1}) \left(\alpha_1 \frac{b_{n+1}}{b_n} + \alpha_2 \frac{b_{n+1}}{b_{n-1}} \right) + \alpha_1 x_n + \alpha_2 x_{n-1} - kt + k$$

$$= k \left(s - x_{n+1} \right) + \alpha_1 \sum_{i=0}^{n} \frac{1}{b_i} + \alpha_2 \sum_{i=0}^{n-1} \frac{1}{b_i} - kt + k$$

$$= ks - k \sum_{i=0}^{n+1} \frac{1}{b_i} + \alpha_1 + \sum_{i=1}^{n} \left(\alpha_1 \frac{1}{b_i} + \alpha_2 \frac{1}{b_{i-1}} \right) - kt + k$$

$$= ks - k \frac{1}{b_0} - k \frac{1}{b_1} - kt + k + \alpha_1$$

$$= k(s - t).$$

Thus, f is (α_1, α_2)-lipschitzian for the constant k.
For each $n \geq 0$, we have $f(x_n) = x_{n-1}$, and

$$f^n(x_n) - f^n(x_{n-1}) = x_0 - x_{-1} = 1 = b_n \cdot \frac{1}{b_n} = b_n (x_n - x_{n-1}).$$

This implies that $k(f^n) \geq b_n$. Finally, by theorem 4.2, $k(f^n) = b_n$.

$\square$

Since the half-line $[0, \infty)$ can be isometrically embedded into every normed space $X \neq \{0\}$, we get the following theorem:

Theorem 4.9 *Let C be a subset of a normed space $(X, \|\cdot\|)$. If C contains a half-line, then for every multi-index $\alpha = (\alpha_1, \alpha_2)$ and $k \in (0, 1]$, there exists a mapping $T : C \to C$, which is (α_1, α_2)-lipschitzian for the constant k and $k(T^n) = b_n$ for $n = 1, 2, \ldots$.*

Proof. Let $u \in C$ and $v \in S_X$ be such that $\{u + tv : t \geq 0\} \subset C$. For a multi-index $\alpha = (\alpha_1, \alpha_2)$ and $k \in (0, 1]$, we define a mapping $T : C \to C$ by putting for each $x \in C$

$$Tx = u + f(\|x - u\|)v,$$

where f is the function as in lemma 4.8.

We claim that the mapping T is (α_1, α_2)-lipschitzian for the constant k and $k(T^n) = b_n$ for $n = 1, 2, \ldots$. Indeed, since f is (α_1, α_2)-lipschitzian for the constant k, this implies that

$$\alpha_1 \|Tx - Ty\| + \alpha_2 \|T^2 x - T^2 y\|$$

$$= \alpha_1 \left| f(\|x - u\|) - f(\|y - u\|) \right| + \alpha_2 \left| f(\|f(\|x - u\|)v\|) - f(\|f(\|y - u\|)v\|) \right|$$

$$= \alpha_1 \left| f(\|x - u\|) - f(\|y - u\|) \right| + \alpha_2 \left| f^2(\|x - u\|) - f^2(\|y - u\|) \right|$$

$$\leq k \left| \|x - u\| - \|y - u\| \right|$$

$$\leq k \|x - y\|$$

for all $x, y \in C$.

Now, it is enough to notice that, for $x = u + x_n v$ and $y = u + x_{n-1} v$, $n \geq 1$, with $\{x_n\}_{n \geq 0}$ as in lemma 4.8, we obtain

$$\begin{aligned} \|T^n x - T^n y\| &= f^n(x_n) - f^n(x_{n-1}) \\ &= b_n (x_n - x_{n-1}) \\ &= b_n \|x - y\|. \end{aligned}$$

Thus, $k(T^n) \geq b_n$ for $n = 1, 2, \ldots$. Finally, by theorem 4.2, we get the conclusion.

$\square$

Corollary 4.7 *Let C be a subset of a normed space $(X, \|\cdot\|)$. If C contains a half-line, then for each multi-index $\alpha = (\alpha_1, \alpha_2)$ and $k \in (0, 1]$, there exists a mapping $T : C \to C$, which is (α_1, α_2)-lipschitzian for the constant k and $k_0(T) = 2k/(\alpha_1 + \sqrt{\Delta})$, where Δ is as in the theorem 4.7.*

Corollary 4.8 *Let C be a subset of a normed space $(X, \|\cdot\|)$. If C contains a half-line, then for each multi-index $\alpha = (\alpha_1, \alpha_2)$, there exists a mapping $T : C \to C$, which is (α_1, α_2)-nonexpansive with $k_\infty(T) = 1 + \alpha_2$.*

Consider now any subset C of X that contains a nontrivial *segment*; that is, there exist two points $u, v \in C$ such that

$$[u, v] := \{(1 - t)u + tv : t \in [0, 1]\} \subset C.$$

Let f be the function as in lemma 4.8 with $k = 1$. Define the function $g : [0, \infty) \to [0, \infty)$ by

$$g(t) = \begin{cases} t & \text{if } t \in [0, 1], \\ f(t - 1) + 1, & \text{if } t \geq 1. \end{cases}$$

In view of lemma 4.8, we conclude that g is (α_1, α_2)-nonexpansive and

$$k(g^n) = \frac{1 + \alpha_2}{1 + (-1)^n \alpha_2^{n+1}} \quad \text{for } n \geq 1.$$

In the next step, we use the function g to construct a family of functions $g_N : [0, \infty) \to [0, \infty)$, $N \geq 1$, as follows (see figure 4.13):

$$g_N(t) = \begin{cases} g(t) & \text{if } t \in [0, x_N + 1], \\ x_{N-1} + 1 & \text{if } t \in [x_N + 1, \infty). \end{cases}$$

It is easy to check that, for every $N \geq 1$, g_N is (α_1, α_2)-nonexpansive. Indeed, if $s, t \in [0, x_N + 1]$, then it follows from the fact that function g is (α_1, α_2)-nonexpansive. If $s \geq x_N + 1$ and $t \geq x_N + 1$, then it is enough to notice that $g_N(s) = g_N(t) = x_{N-1} + 1$. If $s \geq x_N + 1$ and $t \in [0, x_N + 1]$, then

$$\alpha_1 \left(g_N(s) - g_N(t) \right) + \alpha_2 \left(g_N^2(s) - g_N^2(t) \right)$$

$$= \alpha_1 \left(g(x_N + 1) - g(t) \right) + \alpha_2 \left(g^2(x_N + 1) - g^2(t) \right)$$

$$\leq x_N + 1 - t$$

$$\leq s - t.$$

Moreover,

$$k(g_N^n) = k(g^n) = \frac{1 + \alpha_2}{1 + (-1)^n \alpha_2^{n+1}}$$

for $n = 1, \ldots, N$ and

$$k(g_N^n) = 1$$

for $n \geq N + 1$.

Now we are ready to prove the following:

Theorem 4.10 *Let C be a subset of a normed space $(X, \| \cdot \|)$. If C contains a nontrivial segment, then for each multi-index $\alpha = (\alpha_1, \alpha_2)$ and $N \geq 1$, there exists an (α_1, α_2)-nonexpansive mapping $T : C \to C$, with*

$$k(T^n) = \frac{1 + \alpha_2}{1 + (-1)^n \alpha_2^{n+1}}$$

for $n = 1, \ldots, N$ and

$$k(T^n) = 1$$

for $n \geq N + 1$.

Proof. Let $u, v \in C$ be such that $[u, v] \subset C$. Let $\alpha = (\alpha_1, \alpha_2)$, $N \geq 1$ and g_N be as above. The reader can easily verify that the mapping $T : C \to C$ defined by

$$Tx = u + \frac{1}{1 + x_N} g_N \left((1 + x_N) \cdot \frac{\| x - u \|}{\| v - u \|} \right) (v - u)$$

satisfies our claim.

$\square$

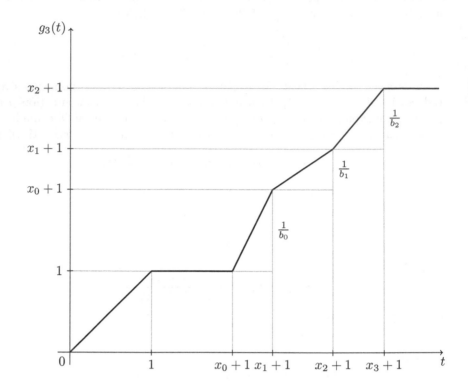

FIGURE 4.13: A graph of function g_3.

In the case of multi-index $\alpha = (\alpha_1, \ldots, \alpha_n)$, it is enough to use the sequence $\{b_m\}$ defined as in theorem 4.3. We leave to the reader verification of the following theorem.

Theorem 4.11 *Let C be a subset of a normed space $(X, \|\cdot\|)$. If C contains a nontrivial segment, then for each multi-index $\alpha = (\alpha_1, \ldots, \alpha_n)$ and $N \geq 1$, there exists an α-nonexpansive mapping $T : C \to C$, with*

$$k(T^m) = b_m$$

for $m = 1, \ldots, N$ and

$$k(T^m) = 1$$

for $m \geq N + 1$.

Consequently, if C is any bounded, closed and convex subset of a Banach space $(X, \|\cdot\|)$, with $\operatorname{diam}(C) > 0$, then, for any multi-index $\alpha = (\alpha_1, \ldots, \alpha_n)$, the class of α-nonexpansive mappings is strictly wider than the class of nonexpansive mappings!

Chapter 5

Subclasses determined by p-averages

Now, we deal with a condition that instead of arithmetical mean in the definition of α-lipschitzian mapping, uses average of order $p \geq 1$.

5.1 Basic definitions and observations

Let (M, ρ) be a metric space, $\alpha = (\alpha_1, \ldots, \alpha_n)$ be a multi-index and $p \geq 1$.

Definition 5.1 *A mapping $T : M \to M$ is said to be (α, p)-lipschitzian for the constant $k \geq 0$ (α and p as above) if, for every $x, y \in M$, we have*

$$\left(\sum_{i=1}^{n} \alpha_i \rho(T^i x, T^i y)^p \right)^{\frac{1}{p}} \leq k\rho(x, y). \tag{5.1}$$

In fact, the above condition was suggested by Goebel and Japón Pineda in [33].

If (5.1) is satisfied with $p = 1$, then we get a definition of α-lipschitzian mappings. For a given α, p, and k, we will denote the class of all (α, p)-lipschitzian mappings $T : M \to M$ for the constant k by $L(\alpha, p, k)$ and if $p = 1$, then we usually write $L(\alpha, k)$ instead of $L(\alpha, 1, k)$. The smallest constant k for which (5.1) holds is denoted by $k(\alpha, p, T)$. If (5.1) is satisfied with $k = 1$, then we say that T is (α, p)-*nonexpansive*, and if (5.1) is satisfied with $k < 1$, then T is called (α, p)-*contraction*.

Let us list some basic properties of (α, p)-lipschitzian mappings.

- Each (α, p)-lipschitzian mapping T is also lipschitzian and

$$k(T) \leq \frac{k(\alpha, p, T)}{\alpha_1^{1/p}}.$$

- For $i = 1, \ldots, n$, we have

$$k(T^i) \leq \min \left[k(T)^i, \frac{k(\alpha, p, T)}{\alpha_i^{1/p}} \right],$$

provided $\alpha_i > 0$.

- If T is lipschitzian, then, for any multi-index $\alpha = (\alpha_1, \ldots, \alpha_n)$ and $p \geq 1$, the mapping T is (α, p)-lipschitzian with mean Lipschitz constant

$$k(\alpha, p, T) \leq \left(\sum_{i=1}^{n} \alpha_i k(T^i)^p \right)^{\frac{1}{p}}.$$

- Each class $L(\alpha, p, k)$ contains all the lipschitzian mappings T such that

$$\left(\sum_{i=1}^{n} \alpha_i k(T^i)^p \right)^{\frac{1}{p}} \leq k.$$

In particular, using the fact that $k(T^i) \leq k(T)^i$, each class $L(\alpha, p, k)$ contains all lipschitzian mappings T such that

$$\left(\sum_{i=1}^{n} \alpha_i k(T)^{ip} \right)^{\frac{1}{p}} \leq k. \tag{5.2}$$

- If T is uniformly lipschitzian with $\sup \left\{ k(T^i) : i = 1, 2, \ldots \right\} \leq k$, then for any α and $p \geq 1$, T is (α, p)-lipschitzian with $k(\alpha, p, T) \leq k$.

- Let $L_u(k)$ denote the class of all uniformly k-lipschitzian mappings. By $\mathfrak{I}$, we denote the set of all multi-indexes α; that is,

$$\mathfrak{I} = \left\{ (\alpha_1, \ldots, \alpha_n) : \alpha_1 > 0, \alpha_n > 0, \alpha_i \geq 0, \sum_{i=1}^{n} \alpha_i = 1, n \geq 1 \right\}.$$

One can notice that, for any $p \geq 1$, we have

$$L_u(k) = \bigcap_{\alpha \in \mathfrak{I}} L(\alpha, p, k). \tag{5.3}$$

Indeed, let us assume that $T \in \bigcap_{\alpha \in \mathfrak{I}} L(\alpha, p, k)$ and $T \notin L_u(k)$. Let m be the smallest natural number for which there exist points $u, v \in M$, $u \neq v$, such that

$$\rho(T^m u, T^m v) > k\rho(u, v)$$

or equivalently

$$\rho(T^m u, T^m v)^p > k^p \rho(u, v)^p \tag{5.4}$$

for every $p \geq 1$.

If $m = 1$, then for the multi-index $\alpha = (1)$ of length $n = 1$, T is not (α, p)-lipschitzian for the constant k.

If $m = 2$, then we set $\alpha = (\alpha_1, \alpha_2)$ with

$$\alpha_1 = \frac{\rho(T^2 u, T^2 v)^p - k^p \rho(u, v)^p}{2\rho(T^2 u, T^2 v)^p}$$

and

$$\alpha_2 = \frac{\rho(T^2 u, T^2 v)^p + k^p \rho(u,v)^p}{2\rho(T^2 u, T^2 v)^p}.$$

Obviously, α is a multi-index of length $n = 2$, and in view of (5.4), we get

$$\alpha_1 \rho(Tu, Tv)^p + \alpha_2 \rho(T^2 u, T^2 v)^p > \frac{\rho(T^2 u, T^2 v)^p + k^p \rho(u,v)^p}{2} > k^p \rho(u,v)^p.$$

In general, if $m \geq 3$, then we define a multi-index $\alpha = (\alpha_1, \ldots, \alpha_m)$ of length $n = m$ by

$$\alpha_i = \begin{cases} \frac{\rho(T^m u, T^m v)^p - k^p \rho(u,v)^p}{2\rho(T^m u, T^m v)^p} & \text{for } i = 1, \\ 0 & \text{for } i = 2, \ldots, m-1, \\ \frac{\rho(T^m u, T^m v)^p + k^p \rho(u,v)^p}{2\rho(T^m u, T^m v)^p} & \text{for } i = m. \end{cases}$$

Further, using (5.4), we get

$$
\begin{aligned}
\sum_{i=1}^{m} \alpha_i \rho(T^i u, T^i v)^p &= \frac{\rho(T^m u, T^m v)^p - k^p \rho(u,v)^p}{2\rho(T^m u, T^m v)^p} \rho(Tu, Tv)^p \\
&\quad + \frac{\rho(T^m u, T^m v)^p + k^p \rho(u,v)^p}{2\rho(T^m u, T^m v)^p} \rho(T^m u, T^m v)^p \\
&> \frac{\rho(T^m u, T^m v)^p + k^p \rho(u,v)^p}{2} \\
&> k^p \rho(u,v)^p.
\end{aligned}
$$

All the above cases contradict with the assumption that, for every α, T is (α, p)-lipschitzian for the constant k. Thus, T is uniformly k-lipschitzian.

Inclusion in opposite direction follows from previous items.

- It is clear that for a given p and q, $1 \leq q < p$, every (α, p)-lipschitzian mapping is (α, q)-lipschitzian for the same constant k. Indeed, if $T \in L(\alpha, p, k)$, then, using the Hölder inequality, we obtain

$$
\begin{aligned}
\left(\sum_{i=1}^{n} \alpha_i \rho(T^i x, T^i y)^q \right)^{\frac{1}{q}} &= \left(\sum_{i=1}^{n} \alpha_i^{\frac{p-q}{p}} \alpha_i^{\frac{q}{p}} \rho(T^i x, T^i y)^q \right)^{\frac{1}{q}} \\
&\leq \left[\left(\sum_{i=1}^{n} \alpha_i \right)^{\frac{p-q}{p}} \left(\sum_{i=1}^{n} \alpha_i \rho(T^i x, T^i y)^p \right)^{\frac{q}{p}} \right]^{\frac{1}{q}} \\
&= \left(\sum_{i=1}^{n} \alpha_i \rho(T^i x, T^i y)^p \right)^{\frac{1}{p}} \\
&\leq k\rho(x, y).
\end{aligned}
$$

This means that $L(\alpha, p, k) \subset L(\alpha, q, k)$ for $p > q$. In particular, for every $p > 1$, $L(\alpha, p, k) \subset L(\alpha, k)$.

Inclusion in opposite direction does not hold. To illustrate this, let us consider the following example.

As a metric space (M, ρ), we take ℓ_1 with a metric inherited from the standard norm. Let $T : \ell_1 \to \ell_1$ be a linear mapping given by

$$Tx = T(x_1, x_2, \dots) = \left(2x_2, \frac{2}{3}x_3, x_4, x_5, \dots \right).$$

Then,

$$T^2 x = T^2(x_1, x_2, \dots) = \left(\frac{4}{3}x_3, \frac{2}{3}x_4, x_5, x_6, \dots \right)$$

and

$$\frac{1}{2} \|Tx\| + \frac{1}{2} \|T^2 x\| = |x_2| + |x_3| + \frac{5}{6}|x_4| + \sum_{i=5}^{\infty} |x_i| \le \|x\|.$$

Thus, T is $(1/2, 1/2)$-nonexpansive; that is, $T \in L\left((1/2, 1/2), 1\right)$. On the other hand, for every $p > 1$, T is not $((1/2, 1/2), p)$-nonexpansive. Indeed, for $x = e_3 = (0, 0, 1, 0, \dots, 0, \dots) \in \ell_1$, we get

$$Te_3 = \left(0, \frac{2}{3}, 0, 0, \dots \right)$$

and

$$T^2 e_3 = \left(\frac{4}{3}, 0, 0, \dots \right).$$

Then,

$$\frac{1}{2} \|Te_3\|^p + \frac{1}{2} \|T^2 e_3\|^p = \frac{1}{2} \cdot \left(\frac{2}{3} \right)^p + \frac{1}{2} \cdot \left(\frac{4}{3} \right)^p > 1 = \|e_3\|^p.$$

- For each multi-index $\alpha = (\alpha_1, \dots, \alpha_n)$, we have

$$L(1) = \bigcap_{p \ge 1} L(\alpha, p, 1).$$

The case with $n = 1$ is trivial. Let $n \ge 2$. Obviously, each class $L(\alpha, p, 1)$ contains all nonexpansive mappings. On the other hand, let us assume that a mapping $T : M \to M$ is in $L(\alpha, p, 1)$ for each $p \ge 1$, and there exist points $u, v \in M$, $u \ne v$, such that

$$\rho(Tu, Tv) > \rho(u, v).$$

For such $u, v \in M$, we take p such that

$$p > \max\left\{1, \frac{\ln(\alpha_1)}{\ln(\rho(u,v)) - \ln(\rho(Tu, Tv))}\right\}.$$

Then,

$$\left(\sum_{i=1}^{n} \alpha_i \rho(T^i u, T^i v)^p\right)^{\frac{1}{p}} \geq \alpha_1^{1/p} \rho(Tu, Tv)$$

$$> \alpha_1^{\frac{\ln(\rho(u,v)/\rho(Tu,Tv))}{\ln(\alpha_1)}} \rho(Tu, Tv)$$

$$= \rho(u,v).$$

By contradiction, we get that T must be nonexpansive.

5.2   A bound for $k(T^n)$, $k_\infty(T)$, and $k_0(T)$

The condition (5.1) determines subclasses $L(\alpha, p, k)$ of class $L(\alpha, k)$. It is natural to expect that, for every such subclass, we can obtain better bounds for $k(T^n)$, $k_0(T)$, and $k_\infty(T)$. Indeed, all results from the previous chapter can be extended easily for subclasses $L(\alpha, p, k)$, and proofs carry on with only minor technical changes.

We begin by listing results devoted to the simple case of multi-index α of length $n = 2$:

Theorem 5.1 *Let* $T : M \to M$ *be* $((\alpha_1, \alpha_2), p)$-*lipschitzian mapping for the constant* $k > 0$. *Then, for* $n \geq 0$, *we have*

$$k(T^n) \leq b_n^{1/p},$$

where

$$b_0 = 1, \quad b_1 = \frac{k^p}{\alpha_1}, \quad \text{and} \quad b_{n+2} = \frac{k^p}{\alpha_1 b_{n+1}^{-1} + \alpha_2 b_n^{-1}}.$$

Proof. It is enough to repeat the proof of theorem 4.2 by putting $\rho(x,y)^p$, $\rho(T^i x, T^i y)^p$, and k^p instead of $\rho(x,y)$, $\rho(T^i x, T^i y)$, and k, respectively.

$\square$

Lemma 5.1 *Let* $\{b_n\}_{n=0}^{\infty}$ *be as above. Then, for* $n \geq 0$, *we have*

$$b_n = k^{pn} \frac{2^{n+1}\sqrt{\Delta}}{\left(\alpha_1 + \sqrt{\Delta}\right)^{n+1} - \left(\alpha_1 - \sqrt{\Delta}\right)^{n+1}},$$

with $\Delta = \alpha_1^2 + 4\alpha_2 k^p$.

Proof. It is enough to put k^p instead of k in the proof of lemma 4.1.

$\square$

Theorem 5.2 *Let* $T : M \to M$ *be* $((\alpha_1, \alpha_2), p)$-*lipschitzian mapping for the constant* $k > 0$. *Then, for* $n \geq 0$, *we have*

$$k(T^n) \leq k^n \left(\frac{2^{n+1}\sqrt{\Delta}}{\left(\alpha_1 + \sqrt{\Delta}\right)^{n+1} - \left(\alpha_1 - \sqrt{\Delta}\right)^{n+1}} \right)^{1/p},$$

where $\Delta = \alpha_1^2 + 4\alpha_2 k^p$.

Proof. It is a consequence of theorem 5.1 and lemma 5.1.

$\square$

Corollary 5.1 *Let* $T : M \to M$ *be* $((\alpha_1, \alpha_2), p)$-*lipschitzian mapping for the constant* $k > 0$. *Then,*

$$k_0(T) \leq k \left(\frac{2}{\alpha_1 + \sqrt{\Delta}} \right)^{1/p},$$

with $\Delta = \alpha_1^2 + 4\alpha_2 k^p$.

Corollary 5.2 *Let* $T : M \to M$ *be* $((\alpha_1, \alpha_2), p)$-*nonexpansive. Then, for* $n \geq 0$, *we have*

$$k(T^n) \leq \left(\frac{1 + \alpha_2}{1 + (-1)^n \alpha_2^{n+1}} \right)^{1/p} \tag{5.5}$$

and

$$k_\infty(T) \leq (1 + \alpha_2)^{1/p}. \tag{5.6}$$

Let us pass now to the general case when the length of multi-index α is equal to n. Then, evaluations presented in theorems 4.3, 4.4, and 4.8, and in corollary 4.6 can be easily extended to the case of (α, p)-lipschitzian mappings. Indeed, it is enough to observe that the condition (5.1) is equivalent to the following:

$$\sum_{i=1}^{n} \alpha_i \rho(T^i x, T^i y)^p \leq k^p \rho(x, y)^p,$$

and then repeat the proofs of the above-mentioned theorems by putting $\rho(x, y)^p$, $\rho(T^i x, T^i y)^p$, and k^p instead of $\rho(x, y)$, $\rho(T^i x, T^i y)$, and k, respectively. Consequently, we get:

Theorem 5.3 *Let* $T : M \to M$ *be* $((\alpha_1, \ldots, \alpha_n), p)$*-lipschitzian for the constant* $k > 0$*. Then, for* $n \geq 0$*, we have*

$$k(T^m) \leq b_m^{1/p},$$

where $\{b_m\}_{m \geq 0}$ *is defined as follows:*

$$b_m = \begin{cases} 1 & \text{for} \quad m = 0, \\ \dfrac{k^p}{\sum\limits_{j=1}^{m} \alpha_j b_{m-j}^{-1}} & \text{for} \quad m = 1, \ldots, n, \\ \dfrac{k^p}{\sum\limits_{i=1}^{n} \alpha_i b_{m-i}^{-1}} & \text{for} \quad m = n+1, n+2, \ldots. \end{cases}$$

Theorem 5.4 *If* $T : M \to M$ *is an* $((\alpha_1, \ldots, \alpha_n), p)$*-nonexpansive and* $\{b_m\}_{m \geq 0}$ *is as in theorem 5.3 with* $k = 1$*, then*

$$k_\infty(T) = \limsup_{m \to \infty} k(T^m) \leq \lim_{m \to \infty} b_m^{1/p} = \left(\sum_{j=1}^{n} \left(\sum_{i=j}^{n} \alpha_i \right) \right)^{1/p}$$

$$= (1 + \alpha_2 + 2\alpha_3 + 3\alpha_4 + \cdots + (n-1)\alpha_n)^{1/p}$$

$$= (\alpha_1 + 2\alpha_2 + 3\alpha_3 + \cdots + n\alpha_n)^{1/p}.$$

Let us observe that the term on the right-hand side of the above inequality is always less than $n^{1/p}$.

Theorem 5.5 *If* $T : M \to M$ *is an* $((\alpha_1, \ldots, \alpha_n), p)$*-lipschitzian for the constant* k *and* $\{b_m\}_{m=0}^{\infty}$ *is as in theorem 5.3, then*

$$k_0(T) = \lim_{m \to \infty} [k(T^m)]^{1/m} \leq \lim_{m \to \infty} \left(b_m^{1/p} \right)^{1/m} = \lim_{m \to \infty} \left(b_m^{1/m} \right)^{1/p} = g^{1/p},$$

where g *is the unique positive solution of the equation*

$$\alpha_1 g + \alpha_2 g^2 + \cdots + \alpha_n g^n = k^p.$$

Moreover, an equivalent metric d_T for which any (α, p)-lipschitzian mapping $T : M \to M$ with $\alpha = (\alpha_1, \ldots, \alpha_n)$, $p \geq 1$, and $k > 0$ satisfies the Lipschitz condition with a constant $g^{1/p}$ can be given by

$$d_T(x, y) = \left[\left(\alpha_1 + \alpha_2 g + \cdots + \alpha_n g^{n-1} \right) \rho(x, y)^p \right.$$
$$+ \left(\alpha_2 + \alpha_3 g + \cdots + \alpha_n g^{n-2} \right) \rho(Tx, Ty)^p$$
$$+ \left(\alpha_3 + \alpha_4 g + \cdots + \alpha_n g^{n-3} \right) \rho(T^2 x, T^2 y)^p + \cdots$$
$$\left. + \alpha_n \rho(T^{n-1} x, T^{n-1} y)^p \right]^{1/p}.$$

To show that all the above estimates are sharp, we present the following example.

Example 5.1 *For each $p \geq 1$, let us consider the space $M = \ell_p$ with a metric inherited from the standard norm, $\|x\| = \|(x_1, x_2, \dots)\| = \left(\sum_{i=1}^{\infty} |x_i|^p\right)^{1/p}$. For arbitrary multi-index $\alpha = (\alpha_1, \dots, \alpha_n)$ and $k > 0$, let $\{b_m\}_{m \geq 0}$ be as in theorem 5.3. Let $T : \ell_p \to \ell_p$ be a linear operator defined for each $x = (x_1, x_2, \dots) \in \ell_p$ by*

$$Tx = \left(\frac{b_1^{1/p}}{b_0^{1/p}} x_2, \frac{b_2^{1/p}}{b_1^{1/p}} x_3, \dots, \frac{b_j^{1/p}}{b_{j-1}^{1/p}} x_{j+1}, \dots\right).$$

The reader can verify that T is $((\alpha_1, \dots, \alpha_n), p)$-lipschitzian with the mean Lipschitz constant $k(\alpha, p, T) = k$. Moreover, $k(T^m) = b_m^{1/p}$ for each $m \geq 0$ and

$$k_0(T) = g^{1/p},$$

where g is as in theorem 5.5. If $k = 1$, then T is $((\alpha_1, \dots, \alpha_n), p)$-nonexpansive with $k_0(T) = 1$ and

$$k_\infty(T) = \limsup_{m \to \infty} k(T^m) = \lim_{m \to \infty} b_m^{1/p} = \left(\sum_{j=1}^{n} \left(\sum_{i=j}^{n} \alpha_i\right)\right)^{1/p}$$

$$= (1 + \alpha_2 + 2\alpha_3 + 3\alpha_4 + \dots + (n-1)\alpha_n)^{1/p}$$

$$= (\alpha_1 + 2\alpha_2 + 3\alpha_3 + \dots + n\alpha_n)^{1/p}.$$

The above mapping can be modified to the $((\alpha_1, \dots, \alpha_n), p)$-nonexpansive mapping $T : B_{\ell_p} \to B_{\ell_p}$, with

$$k_\infty(T) = (1 + \alpha_2 + 2\alpha_3 + \dots + (n-1)\alpha_n)^{1/p}.$$

For clarity and simplicity of arguments, we present it in the case of multi-index α of length $n = 2$.

Example 5.2 *For our construction, we shall need a cut function $\tau : [0, 1] \to [0, 1]$ given by (see figure 5.1).*

$$\tau(t) = \begin{cases} \frac{1}{\alpha_1} t - \frac{\alpha_2^n}{\alpha_1 (2 - \alpha_2)^{n-1}} & \text{if } \frac{\alpha_2^n}{(2-\alpha_2)^{n-1}} \leq t \leq \frac{\alpha_2^{n-1}}{(2-\alpha_2)^{n-1}}, n = 1, 2, \dots, \\ -\frac{1}{\alpha_1} t + \frac{\alpha_2^n}{\alpha_1 (2-\alpha_2)^{n-1}} & \text{if } \frac{\alpha_2^n}{(2-\alpha_2)^n} \leq t \leq \frac{\alpha_2^n}{(2-\alpha_2)^{n-1}}, n = 1, 2, \dots, \\ 0 & \text{if } t = 0. \end{cases}$$

Obviously, for all $s, t \in [0, 1]$, we have

$$|\tau(s) - \tau(t)| \leq \frac{1}{\alpha_1} |s - t|$$

and

$$|\tau(t)| = \tau(t) \leq |t| = t.$$

Fix $p \geq 1$. Then, for any $\alpha = (\alpha_1, \alpha_2)$, we define a mapping $T : B_{\ell_p} \to B_{\ell_p}$ by

$$T(x_i)_{i \geq 1} = \left(\tau \left(\alpha_1 \left(\frac{1 - (-\alpha_2)^i}{1 - (-\alpha_2)^{i+1}} \right)^{1/p} |x_{i+1}| \right) \right)_{i \geq 1}.$$

The reader can verify that

$$k(\alpha, p, T) = 1$$

and

$$k(T^n) = \left(\frac{1 + \alpha_2}{1 + (-1)^n \alpha_2^{n+1}} \right)^{1/p}$$

for each $n \geq 1$. Consequently,

$$k_\infty(T) = (1 + \alpha_2)^{1/p}.$$

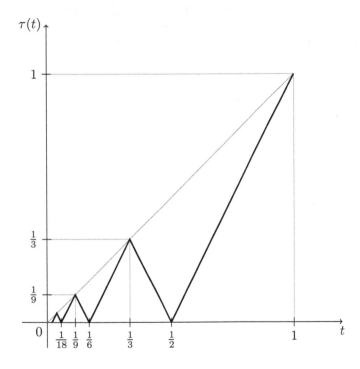

FIGURE 5.1: *A graph of function τ for $\alpha = \left(\frac{1}{2}, \frac{1}{2} \right)$.*

5.3   On the moving p-averages

The result from theorem 4.5 can be easily extended to the moving p-averages. Fix a multi-index $\alpha = (\alpha_1, \ldots, \alpha_n)$ and $p > 0$. Let $d_0, \ldots, d_{n-1}$ be arbitrary nonnegative numbers and d_m be defined for $m \geq n$ by

$$d_m = \left(\sum_{i=1}^{n} \alpha_i d_{m-i}^p \right)^{1/p}$$

or, equivalently,

$$d_m^p = \sum_{i=1}^{n} \alpha_i d_{m-i}^p.$$

Then,

$$\lim_{m \to \infty} d_m = \left(\frac{1}{\sum_{i=1}^{n} \sum_{j=i}^{n} \alpha_j} \sum_{i=1}^{n} \left(\sum_{j=i}^{n} \alpha_j \right) d_{n-i}^p \right)^{1/p}.$$

To see this, it is enough to put b_m^p instead of b_m and d_m^p instead of d_m in the proof of the classical moving average problem presented in the previous chapter.

Chapter 6

Mean contractions

Let us pass now to the class of mean contractions. We have already seen that every such mapping is a classical contraction with respect to some equivalent metric. Actually, as we shall see later, every mean contraction is characterized by the constant $k_0(T)$ as well as $k_\infty(T)$. Moreover, the mean Lipschitz condition provides nice estimates on the rate of convergence of iterates.

6.1   Classical Banach's contractions

To present the just mentioned results, we begin by recalling the Banach's Contraction Mapping Principle and some of its useful application variants.

Let (M, ρ) be a metric space. Let us recall that a mapping $T : M \to M$ is said to be *a contraction* if there exists a constant $k < 1$ such that, for all x, $y \in M$, we have

$$\rho(Tx, Ty) \leq k\rho(x, y).$$

Under the above setting the Banach's Contraction Mapping Principle states:

Theorem 6.1 *Let (M, ρ) be a complete metric space and let $T : M \to M$ be a k-contraction. Then, T has a unique fixed point z. Moreover, for each $x \in M$,*

$$\lim_{n \to \infty} T^n x = z$$

and, in terms of error estimates,

- $\rho(T^n x, z) \leq \frac{1}{1-k} \rho(T^n x, T^{n+1} x)$ *for each $x \in M$ and $n \geq 1$;*

- $\rho(T^n x, z) \leq \frac{k^n}{1-k} \rho(x, Tx)$ *for each $x \in M$ and $n \geq 1$.*

Proof. Since T is a contraction, we conclude that, for each $x \in M$,

$$\rho(x, Tx) \leq (1 - k)^{-1} \left(\rho(x, Tx) - \rho(Tx, T^2 x) \right).$$

Therefore, for any $m, n \in \mathbb{N}$ with $n < m$, we have

$$\rho(T^n x, T^{m+1} x) \leq \sum_{i=n}^{m} \rho(T^i x, T^{i+1} x)$$

$$\leq (1-k)^{-1} \sum_{i=n}^{m} \left(\rho(T^i x, T^{i+1} x) - \rho(T^{i+1} x, T^{i+2} x) \right)$$

$$= (1-k)^{-1} \left(\rho(T^n x, T^{n+1} x) - \rho(T^{m+1} x, T^{m+2} x) \right).$$

In particular, for $n = 1$ and letting $m \to \infty$, we obtain

$$\sum_{i=1}^{\infty} \rho(T^i x, T^{i+1} x) \leq (1-k)^{-1} \rho(Tx, T^2 x) < \infty.$$

This implies that $\{T^n x\}_{n=1}^{\infty}$ is a Cauchy sequence. Since M is complete, there exists a point $z \in M$ such that

$$\lim_{n \to \infty} T^n x = z.$$

By continuity of T, we get

$$z = \lim_{n \to \infty} T^n x = \lim_{n \to \infty} T^{n+1} x = \lim_{n \to \infty} T(T^n x) = Tz.$$

Thus, $z \in \text{Fix}(T)$. Letting $m \to \infty$ in the above proved inequality,

$$\rho(T^n x, T^{m+1} x) \leq (1-k)^{-1} \left(\rho(T^n x, T^{n+1} x) - \rho(T^{m+1} x, T^{m+2} x) \right),$$

we obtain

$$\rho(T^n x, z) \leq (1-k)^{-1} \rho(T^n x, T^{n+1} x),$$

and, consequently,

$$\rho(T^n x, z) \leq \frac{k^n}{1-k} \rho(x, Tx).$$

To show that z is the unique fixed point, let us assume that $Ty = y$ for some $y \in M$. Then,

$$\rho(y, z) = \rho(Ty, Tz) \leq k\rho(y, z).$$

Since $k < 1$, we conclude that $y = z$.

$$\square$$

The simple example $Tx = kx$ for $x \in \mathbb{R}$ shows that the above estimates are sharp. The Banach's theorem fails to hold when $k(T) = 1$. To show this, it is enough to consider a mapping $T : \mathbb{R} \to \mathbb{R}$ defined by $Tx = x + a$, $a \neq 0$.

It is known that the assumption $k < 1$ can be replaced by (see [36], pp. 10–11)

$$\sum_{i=1}^{\infty} k(T^i) < \infty. \tag{6.1}$$

Indeed, for any $x \in M$ and $n, p \in \mathbb{N}$, we have

$$\rho(T^n x, T^{n+p} x) \le \sum_{i=0}^{p-1} \rho(T^{n+i} x, T^{n+i+1} x) \le \left(\sum_{i=n}^{n+p-1} k(T^i) \right) \rho(x, Tx).$$

Using (6.1) we conclude that $\{T^n x\}$ is again a Cauchy sequence. Since M is complete and T is lipschitzian, $\{T^n x\}$ converges to the fixed point z of T. Thus, letting p to go to infinity, we obtain for any fixed n

$$\rho(T^n x, z) \le \left(\sum_{i=n}^{\infty} k(T^i) \right) \rho(x, Tx).$$

In view of (6.1), there is a number m such that $k(T^m) < 1$. Thus, by the Banach's theorem, $T^m : M \to M$ has exactly one fixed point. This implies that z is the unique fixed point of T.

We can characterize lipschitzian mappings that satisfy (6.1). Let us recall that, for the constant $k_0(T)$, we have (see (1.5) and (1.10)):

$$k_0(T) = \lim_{n \to \infty} \sqrt[n]{k(T^n)} = \inf \left\{ \sqrt[n]{k(T^n)} : n = 1, 2, \dots \right\} = \inf k_r(T), \quad (6.2)$$

where infimum is taken over all metrics r equivalent to ρ. Thus, the series (6.1) converges if and only if $k_0(T) < 1$.

We summarize the above observation in the following:

Theorem 6.2 *Let (M, ρ) be a complete metric space. Suppose that $T : M \to M$ is lipschitzian with $k_0(T) < 1$. Then, T has a unique fixed point z, and for each $x \in M$, $\lim_{n \to \infty} T^n x = z$.*

Even if $k_\rho(T) < 1$, it may happen that $k_0(T) < k_\rho(T)$. Then, we obtain better estimate on the rate of convergence as seen in the following:

Theorem 6.3 *Let (M, ρ) be a complete metric space and $T : M \to M$ be a lipschitzian mapping with $k_0(T) < 1$. Then, T has a unique fixed point z and for each $x \in M$ and $\epsilon > 0$,*

$$\rho(T^n x, z) = o((k_0(T) + \epsilon)^n).$$

Proof. Let $\{d_\lambda : \lambda \in (0, 1/k_0(T))\}$ be a family of equivalent metrics defined by (1.7). For fixed $\epsilon > 0$, we take $\delta > 0$ such that $\delta < \min\{\epsilon, 1 - k_0(T)\}$. Then, using (1.9), for $\lambda = 1/(k_0(T) + \delta)$, we have $k_{d_\lambda}(T) \le k_0(T) + \delta < 1$. Thus, by the Banach's Contraction Mapping Principle, there exists the unique point z such that $z = Tz$, and for each $x \in M$, we have

$$\lim_{n \to \infty} d_\lambda(T^n x, z) = 0.$$

Since metric d_λ is equivalent to ρ, we obtain

$$\lim_{n \to \infty} \rho(T^n x, z) = 0.$$

For fixed $x \in M$ and $n, p \in \mathbb{N}$, using (1.8) and (1.9), we have

$$\rho(T^n x, T^{n+p} x) \leq d_\lambda(T^n x, T^{n+p} x)$$

$$\leq \sum_{i=n}^{n+p-1} d_\lambda(T^i x, T^{i+1} x)$$

$$\leq \left[\sum_{i=n}^{n+p-1} (k_0(T) + \delta)^i \right] d_\lambda(x, Tx)$$

$$\leq \frac{(k_0(T) + \delta)^n}{1 - k_0(T) - \delta} d_\lambda(x, Tx)$$

$$\leq \frac{(k_0(T) + \delta)^n}{1 - k_0(T) - \delta} \left[\sum_{i=0}^{\infty} k_\rho(T^i) \left(\frac{1}{k_0(T) + \delta} \right)^i \right] \rho(x, Tx).$$

Letting $p \to \infty$, we obtain

$$\rho(T^n x, z) \leq \frac{(k_0(T) + \delta)^n}{1 - k_0(T) - \delta} \left[\sum_{i=0}^{\infty} k_\rho(T^i) \left(\frac{1}{k_0(T) + \delta} \right)^i \right] \rho(x, Tx)$$

$$= o((k_0(T) + \epsilon)^n).$$

$\square$

An application of the above presented variants of the Banach's Contraction Mapping Principle in the theory of differential equations can be found in [36].

6.2   On characterizations of contractions

Let us recall that $T : M \to M$ is called *mean contraction* if there exist a multi-index $\alpha = (\alpha_1, \dots, \alpha_n)$ and a constant $k < 1$ such that, for all $x, y \in M$, we have

$$\sum_{i=1}^{n} \alpha_i \rho(T^i x, T^i y) \leq k \rho(x, y).$$

The following theorem characterizes the class of all mean contractions:

Theorem 6.4 *Let (M, ρ) be a metric space. If $T : M \to M$ is lipschitzian, then the following conditions are equivalent:*

(i) *T is mean contraction;*

(ii) *$k_0(T) < 1$;*

(iii) *$k_\infty(T) = 0$;*

(iv) $k_\infty(T) < 1$;

(v) *There exists m such that $k(T^m) < 1$;*

(vi) $\sum_{i=1}^\infty k(T^i) < \infty$.

Proof. Equivalence of (ii), (iii), (iv), (v), and (vi) follows directly from (6.2). Thus, it is enough to prove that T is mean contraction if and only if $k_0(T) < 1$.

If T is mean contraction, then, in view of theorem 3.8, there exists an equivalent metric d such that $k_d(T) < 1$. Thus, $k_0(T) < 1$.

If $k_0(T) < 1$, then there exists an equivalent metric d such that $k_d(T) < 1$. Then,

$$\lim_{m\to\infty} k_d(T^m) = 0,$$

and, consequently,

$$\lim_{m\to\infty} k_\rho(T^m) = 0.$$

Let n be the smallest natural number such that $k_\rho(T^n) < 1$. If $n = 1$, that is, $k_\rho(T) < 1$, then for any multi-index α, T is an α-contraction. If $n = 2$, then for a multi-index $\alpha = (\alpha_1, \alpha_2)$ given by

$$\alpha_i = \begin{cases} \frac{1-k_\rho(T^2)}{2k_\rho(T)} & \text{for } i = 1, \\ \frac{k_\rho(T^2)+2k_\rho(T)-1}{2k_\rho(T)} & \text{for } i = 2, \end{cases}$$

we have

$$\alpha_1 \rho(Tx, Ty) + \alpha_2 \rho(T^2 x, T^2 y)$$
$$\leq \left(\alpha_1 k_\rho(T) + \alpha_2 k_\rho(T^2)\right) \rho(x,y)$$
$$\leq \left(\alpha_1 k_\rho(T) + k_\rho(T^2)\right) \rho(x,y)$$
$$= \left(\frac{1 - k_\rho(T^2)}{2} + k_\rho(T^2)\right) \rho(x,y)$$
$$= \frac{1 + k_\rho(T^2)}{2} \rho(x,y).$$

This implies that $k(\alpha, T) \leq (1 + k_\rho(T^2))/2 < 1$. If $n \geq 3$, then for a multi-index $\alpha = (\alpha_1, \dots, \alpha_n)$ given by

$$\alpha_i = \begin{cases} \frac{1-k_\rho(T^n)}{2k_\rho(T)} & \text{for } i = 1, \\ 0 & \text{for } i = 2, \dots, n - 1, \\ \frac{k_\rho(T^n)+2k_\rho(T)-1}{2k_\rho(T)} & \text{for } i = n, \end{cases}$$

the mapping T is an α-contraction.

$\square$

A similar result for mean nonexpansive mappings fails to hold, that is, $k_0(T) = 1$ does not imply that T is α-nonexpansive for some multi-index α (see example 3.6). It is easy to observe that, for any lipschitzian mapping T, we have

$$k_0(T) = 1 \iff k_\infty(T) \in [1, \infty)$$

and

$$k_0(T) > 1 \iff k_\infty(T) = \infty.$$

In view of theorem 6.4, we conclude that a mapping T is a contraction with respect to some equivalent metric if and only if T is mean contraction. In the proof of this theorem, we described how to determine α and k. However, it is clear that the choice of a multi-index α and a constant $k < 1$ is not unique. Even if $k(T) > 1$ and $k(T^2) > 1$, it may happen that for some multi-index $\alpha = (\alpha_1, \alpha_2)$ of length $n = 2$ and $k < 1$, T is in $L(\alpha, k)$:

Example 6.1 *As a metric space (M, ρ), we take ℓ_1 with a metric inherited from the standard norm. Let $T : \ell_1 \to \ell_1$ be as in example 4.2. Then, for $\alpha = (1/2, 1/2)$ and $k = 5/6$, we have $k(T) = 5/3 > 1$ and $k(T^2) = 25/24 > 1$ (see figure 4.4). Hence, despite having the first and the second iterate strictly expansive for some pairs of points in ℓ_1, T can be classified into the class $L((1/2, 1/2), 5/6)$.*

6.3   On the rate of convergence of iterates

The estimate presented in the theorem 6.3 requires knowledge of the behavior of Lipschitz constants for all iterates of the mapping T. In practice, it is hard to determine or to give nice evaluations for them. We may try to remedy this problem using the condition of mean contraction as seen below.

If, for some $\alpha = (\alpha_1, \alpha_2)$ and $k < 1$, T can be classified into the class $L(\alpha, k)$, then it is enough to use one of the following theorems to get a "nice" rate of convergence of iterates:

Theorem 6.5 *Let (M, d) be a complete metric space, and let $T : M \to M$ be an (α_1, α_2)-contraction for the constant $k < 1$. Then, T has a unique fixed point z. Moreover, for each $x \in M$,*

$$\lim_{n \to \infty} T^n x = z,$$

and in terms of error estimates for each $x \in M$ and $n \geq 1$,

- $\rho(T^n x, z) \leq \frac{1}{1-k} \left(\rho(T^n x, T^{n+1} x) + \alpha_2 \rho(T^{n+1} x, T^{n+2} x) \right)$;

- $\rho(T^n x, z) \leq \frac{1}{1-k} \left(k \rho(T^{n-1} x, T^n x) + \alpha_2 \rho(T^n x, T^{n+1} x) \right)$.

Proof. Convergence of the sequence $\{T^n x\}_{n=1}^{\infty}$ to the unique fixed point z follows from theorem 6.4 and the Banach's Contraction Mapping Principle.

Let $n \geq 1$ and $j \geq 4$. By definition of T, we can write a system of inequalities for $i = 1, \ldots, j-2$,

$$\alpha_1 \rho(T^{n+i}x, T^{n+i+1}x) + \alpha_2 \rho(T^{n+i+1}x, T^{n+i+2}x) \leq k\rho(T^{n+i-1}x, T^{n+i}x)$$

and adding both sides, we obtain

$$\alpha_1 \rho(T^{n+1}x, T^{n+2}x) + \sum_{i=n+2}^{n+j-2} \rho(T^i x, T^{i+1}x)$$

$$+\alpha_2 \rho(T^{n+j-1}x, T^{n+j}x) \leq k \sum_{i=n}^{n+j-3} \rho(T^i x, T^{i+1}x).$$

Adding to both sides the number

$$\rho(T^n x, T^{n+1}x) + \alpha_2 \rho(T^{n+1}x, T^{n+2}x) + k\rho(T^{n+j-2}x, T^{n+j-1}x)$$
$$+ (\alpha_1 + k)\,\rho(T^{n+j-1}x, T^{n+j}x),$$

we get

$$\sum_{i=n}^{n+j-1} \rho(T^i x, T^{i+1}x) + k\rho(T^{n+j-1}x, T^{n+j}x)$$

$$+k\rho(T^{n+j-2}x, T^{n+j-1}x) \leq k \sum_{i=n}^{n+j-1} \rho(T^i x, T^{i+1}x)$$
$$+\rho(T^n x, T^{n+1}x)$$
$$+\alpha_2 \rho(T^{n+1}x, T^{n+2}x)$$
$$+\alpha_1 \rho(T^{n+j-1}x, T^{n+j}x).$$

Moving the first term from the right-hand side to the left and the second with the third term from the left-hand side to the right, we obtain

$$(1-k) \sum_{i=n}^{n+j-1} \rho(T^i x, T^{i+1}x) \leq \rho(T^n x, T^{n+1}x) + \alpha_2 \rho(T^{n+1}x, T^{n+2}x)$$

$$-k\rho(T^{n+j-2}x, T^{n+j-1}x)$$
$$+ (\alpha_1 - k)\,\rho(T^{n+j-1}x, T^{n+j}x).$$

Dividing both sides by $(1-k) > 0$, we get

$$\sum_{i=n}^{n+j-1} \rho(T^i x, T^{i+1}x) \leq \frac{1}{1-k}\left(\rho(T^n x, T^{n+1}x) + \alpha_2 \rho(T^{n+1}x, T^{n+2}x)\right)$$

$$-\frac{k}{1-k}\rho(T^{n+j-2}x, T^{n+j-1}x)$$
$$+\frac{\alpha_1 - k}{1-k}\rho(T^{n+j-1}x, T^{n+j}x).$$

Using the triangle inequality and the above estimate, we can write

$$\rho(T^n x, T^{n+j}x) \leq \sum_{i=n}^{n+j-1} \rho(T^i x, T^{i+1}x)$$

$$\leq \frac{1}{1-k}\left(\rho(T^n x, T^{n+1}x) + \alpha_2\rho(T^{n+1}x, T^{n+2}x)\right)$$

$$-\frac{k}{1-k}\rho(T^{n+j-2}x, T^{n+j-1}x) + \frac{\alpha_1 - k}{1-k}\rho(T^{n+j-1}x, T^{n+j}x).$$

Letting $j \to \infty$, we obtain

$$\rho(T^n x, z) \leq \frac{1}{1-k}\left(\rho(T^n x, T^{n+1}x) + \alpha_2\rho(T^{n+1}x, T^{n+2}x)\right).$$

The right-hand side of the above inequality can be rewritten as follows:

$$\frac{1}{1-k}\left(\alpha_1\rho(T^n x, T^{n+1}x) + \alpha_2\rho(T^{n+1}x, T^{n+2}x) + \alpha_2\rho(T^n x, T^{n+1}x)\right),$$

and by definition of T, we get

$$\rho(T^n x, z) \leq \frac{1}{1-k}\left(k\rho(T^{n-1}x, T^n x) + \alpha_2\rho(T^n x, T^{n+1}x)\right).$$

$\square$

The reader can verify that the above estimates are sharp. Hint to the proof: use the function f defined as in lemma 4.8. It is easy to generalize the above theorem to the case of $(\alpha_1, \ldots, \alpha_n)$-contractions, and the proof is similar to the one just presented. We leave it to the reader as an exercise.

Theorem 6.6 *If $T : M \to M$ is α-contraction for the constant k with $\alpha = (\alpha_1, \ldots, \alpha_n)$ and $z \in M$ is a fixed point of T, then for each $\epsilon > 0$ and $x \in M$*

$$\rho(T^n x, z) = o\left((g+\epsilon)^n\right),$$

where g is the unique positive solution of the equation

$$\alpha_1 g + \alpha_2 g^2 + \cdots + \alpha_n g^n = k.$$

Proof. It follows from theorem 6.3 and corollary 4.6.

$\square$

Corollary 6.1 *If $T : M \to M$ is (α_1, α_2)-contraction for the constant k and $z \in M$ is a fixed point of T, then for each $\epsilon > 0$ and $x \in M$,*

$$\rho(T^n x, z) = o\left(\left(\frac{2k}{\alpha_1 + \sqrt{\Delta}} + \epsilon\right)^n\right),$$

with $\Delta = \alpha_1^2 + 4\alpha_2 k$.

Chapter 7

Nonexpansive mappings in Banach space

The metric fixed point theory for nonexpansive mappings has its foundations in works of Browder [17], Göhde [43], and Kirk [48], published in 1965. Currently, it is known that f.p.p. depends deeply on the norm geometry. In the literature, the reader may find a huge collection of conditions put on the space X or the set C that guarantee possessing the fixed point property. It is not our aim to establish all of them. We rather stress those classical as well as very recent results that are sufficient to understand the topic under discussion. An excellent overview of almost all known facts can be found in the handbook [49].

7.1 The asymptotic center technique

One of main tools in the fixed point theory is the **asymptotic center** introduced by Edelstein [26] in 1972.

Let C be a nonempty, closed and convex subset of X. For $x \in C$ and a bounded sequence $\{x_n\} \subset X$, we define *the asymptotic radius of* $\{x_n\}$ *at* x as the number

$$r(x, \{x_n\}) = \limsup_{n \to \infty} \|x - x_n\|. \tag{7.1}$$

If $\{x_n\}$ is fixed, then the formula (7.1) defines a function on C, which has the following properties:

- For all $x \in C$, $r(x, \{x_n\}) = 0 \Leftrightarrow \lim_{n \to \infty} x_n = x$.

- For all $x, y \in C$, $|r(x, \{x_n\}) - r(y, \{x_n\})| \leq \|x - y\|$.

- For all $x, y \in C$, and $\alpha, \beta \geq 0$ with $\alpha + \beta = 1$,

$$r(\alpha x + \beta y, \{x_n\}) \leq \alpha r(x, \{x_n\}) + \beta r(y, \{x_n\}).$$

Therefore, $r(x, \{x_n\})$ is a nonnegative, continuous, and convex function of x. The greatest bound from below,

$$r(C, \{x_n\}) = \inf \{r(y, \{x_n\}) : y \in C\},$$

is called *the asymptotic radius of* $\{x_n\}$ *in* C. The set

$$A\left(C,\{x_n\}\right) = \{y \in C : r(y,\{x_n\}) = r\left(C,\{x_n\}\right)\}$$

is called *the asymptotic center of* $\{x_n\}$ *in* C. The asymptotic center can be defined as an intersection of a descending family of nonempty, bounded, closed and convex subsets of C:

$$A\left(C,\{x_n\}\right) = \bigcap_{\epsilon > 0} \{y \in C : r(y,\{x_n\}) \leq r\left(C,\{x_n\}\right) + \epsilon\}. \qquad (7.2)$$

Hence, the asymptotic center is always closed and convex. It may be empty or nonempty (in particular if C is weakly compact) and consists of one or more points. Structure of such sets depends deeply on the norm geometry of X.

Example 7.1 *Consider the space* $X = \mathbb{R}^2$ *furnished with the norm* $\|(x_1, x_2)\| = \max\{|x_1|, |x_2|\}$. *As a subset* C, *take the closed unit ball* $B = \{(x_1, x_2) : |x_1| \leq 1, |x_2| \leq 1\}$. *Consider the sequence* $z_n = ((-1)^n, 0)$. *Then,*

$$A\left(B,\{z_n\}\right) = \{(0, x_2) : x_2 \in [-1, 1]\}.$$

A connection between the asymptotic center of the sequence and the characteristic of convexity $\epsilon_0(X)$ is presented in the following lemma (see [36], pp. 91)

Lemma 7.1 *If* C *is a nonempty, closed, and convex subset of a Banach space* X, *then, for each bounded sequence* $\{x_n\}$ *in* X, *we have*

$$diam\left(A\left(C,\{x_n\}\right)\right) \leq \epsilon_0\left(X\right) r\left(C,\{x_n\}\right).$$

Proof. If $A\left(C,\{x_n\}\right)$ is empty or consists of exactly one point, then there is nothing to prove. Suppose that $d := \operatorname{diam}\left(A\left(C,\{x_n\}\right)\right) > 0$. For each $\epsilon \in (0, d)$, there exist points $y, z \in A\left(C,\{x_n\}\right)$ such that $\|y - z\| \geq d - \epsilon$. Since $A\left(C,\{x_n\}\right)$ is convex, $\frac{y+z}{2} \in A\left(C,\{x_n\}\right)$. Denote $r := r\left(C,\{x_n\}\right)$. Then $r > 0$ and

$$\left. \begin{array}{l} \limsup\limits_{n\to\infty} \|x_n - y\| = r \\ \limsup\limits_{n\to\infty} \|x_n - z\| = r \end{array} \right\} \Longrightarrow$$

$$\Longrightarrow r = \limsup_{n\to\infty} \left\| x_n - \frac{y+z}{2} \right\| \leq \left(1 - \delta_X\left(\frac{d-\epsilon}{r}\right)\right) r.$$

Consequently,

$$\frac{d}{r} \leq \epsilon_0\left(X\right).$$

$\square$

Observe that in example 7.1 we have $r\left(B,\{z_n\}\right) = 1$, $\epsilon_0(X) = 2$, and $\text{diam}\left(A\left(B,\{z_n\}\right)\right) = 2$.

If X is uniformly convex, then the asymptotic center $A\left(C,\{x_n\}\right)$ is nonempty as an intersection of weakly compact sets (uniform convexity implies reflexivity). By lemma 7.1, it consists of exactly one point ($\epsilon_0\left(X\right) = 0$).

The preceding observation may be used in the proof of the well-known theorem of Browder [17] and Göhde [43].

Theorem 7.1 (Browder and Göhde, 1965) *Bounded, closed and convex subsets of uniformly convex Banach X space have the fixed point property for nonexpansive mappings.*

Proof. Let C be a bounded, closed and convex subset of X. Suppose $\{x_n\}$ is an approximate fixed point sequence for nonexpansive mapping $T : C \to C$; that is,

$$\lim_{n\to\infty} \|x_n - Tx_n\| = 0.$$

Let $r = r\left(C,\{x_n\}\right)$ and $\{z\} = A(C,\{x_n\})$. Observe that

$$\|Tz - x_n\| \leq \|Tz - Tx_n\| + \|Tx_n - x_n\|$$
$$\leq \|z - x_n\| + \|Tx_n - x_n\|.$$

Hence, putting limsup to both sides of the above inequality, we obtain

$$r(Tz,\{x_n\}) \leq r(z,\{x_n\}) = r.$$

This implies that $Tz \in A\left(C,\{x_n\}\right)$. Thus, $Tz = z$.

$\square$

Similarly, we can prove the above theorem using the sequence of iterates of a given point $x_0 \in C$, $\{T^n(x_0)\}$.

If C is compact and $\{x_n\}$ is an approximate fixed point sequence for $T : C \to C$, then $\{x_n\}$ may fail to converge. Nevertheless, by compactness of C, we can always choose a convergent subsequence $\{x_{n_k}\}$ and then $\lim_{k\to\infty} x_{n_k} = z = Tz$. Consider, for instance, the mapping $T : [-1,1] \to [-1,1]$ given by $Tx = x$ and the sequence $x_n = (-1)^n$. In the general case of bounded, closed and convex subset C it does not hold.

Example 7.2 *Fix $p \in [1,\infty)$. Let $X = \ell_p$ with the standard norm,*

$$\|x\| = \|(x_1, x_2,\dots)\| = \left(\sum_{i=1}^{\infty} |x_i|^p\right)^{\frac{1}{p}}.$$

Let $\{\xi_i\}_{i=1}^{\infty}$ be a sequence of real numbers in $(0,1)$ such that $\lim_{i\to\infty} \xi_i = 1$. Define the mapping $T : B_{\ell_p} \to B_{\ell_p}$ by

$$Tx = T\left(x_1, x_2, x_3, \dots\right) = (\xi_1 x_1, \xi_2 x_2, \xi_3 x_3, \dots).$$

It is easy to see that $Fix(T) = \{(0,0,\dots,0,\dots)\}$. If e_n denotes n-th vector of standard basis, then

$$Te_n = (\underbrace{0,\dots,0}_{n-1}, \xi_n, 0, \dots, 0, \dots)$$

and

$$\lim_{n\to\infty} \|Te_n - e_n\| = \lim_{n\to\infty} \|(0,\dots,0,\xi_n - 1, 0,\dots,0,\dots)\| = \lim_{n\to\infty} |\xi_n - 1| = 0.$$

Hence, $\{e_n\}$ is an approximate sequence for T. However, $\{e_n\}$ does not contain any convergent subsequence.

Observe that, for $p \in (1,\infty)$, the approximate sequence $\{e_n\}$ of mapping T from the above example is weakly convergent, and its weak limit is a fixed point of T,

$$w - \lim_{n\to\infty} e_n = (0,0,\dots,0,\dots) = T(0,0,\dots,0,\dots).$$

If $p = 1$, then $\{e_n\}$ is weakly* convergent to the origin in $\ell_1 = c_0^*$. It is not a coincidence. Recall that a bounded sequence in ℓ_p with $p \in (1,\infty)$ (or in $\ell_1 = c_0^*$) is weak convergent (weak* convergent, respectively) if and only if it converges coordinate-wise. Consider first the case of space $\ell_1 = c_0^*$.

Lemma 7.2 *If $\{x^n\}_{n=1}^\infty$ is a bounded sequence in ℓ_1, which converges coordinate-wise to x, then for any $y \in \ell_1$,*

$$r(y, \{x^n\}) = r(x, \{x^n\}) + \|x - y\|.$$

Proof. Since

$$\|x^n - y\| \le \|x^n - x\| + \|x - y\|,$$

we immediately obtain

$$r(y, \{x^n\}) \le r(x, \{x^n\}) + \|x - y\|.$$

Consider now the family of natural projections P_i of ℓ_1 onto the subspace span $\{e_1, \dots, e_i\}$, that is, for $i = 1, 2, \dots$ and $x = (x_1, x_2, \dots) \in \ell_1$, we put

$$P_i(x_1, x_2, \dots) = (x_1, x_2, \dots, x_i, 0, \dots, 0, \dots).$$

Let $Q_i = I - P_i$, where I denotes identity. Then

$$Q_i(x_1, x_2, \dots) = (\underbrace{0, \dots, 0}_{i}, x_{i+1}, x_{i+2}, \dots).$$

For any $i = 1, 2, \dots,$

$$r(x, \{x^n\}) = \limsup_{n\to\infty} \|x^n - x\| = \limsup_{n\to\infty} \|Q_i(x^n - x)\|.$$

Moreover,

$$\|x - y\| = \lim_{i \to \infty} \left(\lim_{n \to \infty} \|P_i (x^n - y)\| \right)$$

and

$$\lim_{i \to \infty} \|Q_i (x - y)\| = 0.$$

Observe that

$$\|P_i (x^n - y)\| + \|Q_i (x^n - x)\| \leq \|x^n - y\| + \|Q_i (x - y)\|.$$

Letting $k \to \infty$ and then $i \to \infty$, we finally obtain

$$r (y, \{x^n\}) \geq r (x, \{x^n\}) + \|x - y\|.$$

$\square$

Similarly, if $\{x^n\}_{n=1}^{\infty}$ is a bounded sequence in ℓ_p with $p \in (1, \infty)$, which converges coordinate-wise to x, then for any $y \in \ell_p$,

$$(r (y, \{x^n\}))^p = (r (x, \{x^n\}))^p + \|x - y\|^p. \tag{7.3}$$

Let C be a bounded closed and convex subset of ℓ_p with $1 < p < \infty$ and let $T : C \to C$ be a nonexpansive mapping. Consider any approximate fixed point sequence $\{x^n\} \subset C$ of T. Since ℓ_p is reflexive, the set C is weakly compact. Hence, in view of Eberlein-Smulian Theorem, there exists a subsequence $\{x^{n_k}\}$ of $\{x^n\}$, which converges weakly to a point $x \in C$. Using (7.3), for any $y \in C$, $y \neq x$, we have $r (y, \{x^{n_k}\}) > r (x, \{x^{n_k}\})$ and, consequently, $A (\{x^{n_k}\}, C) = \{x\}$. Following the proof of theorem 7.1, we also conclude that $Tx \in A (\{x^{n_k}\}, C)$. Hence, it must be the case $Tx = x$.

Let us return now to the case of ℓ_1 space. Suppose C is a weakly* compact and convex subset of $\ell_1 = c_0^*$ and $T : C \to C$ is a nonexpansive mapping. Then there exists an approximate fixed point sequence $\{x^n\}$ of T. Recall that the relative weak* topology on bounded sets in $\ell_1 = c_0^*$ is metrizable because c_0 is separable. Hence, we can choose subsequence $\{x^{n_k}\}$, which converges weakly* to a point $x \in C$ (observe that similar reasoning could be applied in the previous case of ℓ_p spaces with $1 < p < \infty$). In view of lemma 7.2, $r (y, \{x^{n_k}\}) > r (x, \{x^{n_k}\})$ for any $y \in C$ such that $y \neq x$. Thus, $A (\{x^{n_k}\}, C) = \{x\}$. Obviously, x is a fixed point of T.

We shall summarize the above observation in the following theorem [47]:

Theorem 7.2 (Karlovitz, 1976) *Weakly* compact and convex subsets of ℓ_1 considered as a dual of c_0 have the f.p.p. for nonexpansive mappings.*

In view of Alaoglu's Theorem and the above result, we conclude that all closed balls in ℓ_1 have the f.p.p. for nonexpansive mappings.

Also, weakly compact and convex subsets of ℓ_1 have the fixed point property. It is a consequence of Schur's Lemma, which states that a bounded sequence in ℓ_1 is weakly convergent if and only if it is norm convergent. Hence, in view of Eberlein-Smulian Theorem, a subset C is weakly compact if and only if it is compact in the norm topology. Finally, by compactness of C, for any nonexpansive mapping $T : C \to C$, there exists a point $z \in C$ such that $\|z - Tz\| = \inf\{\|x - Tx\| : x \in C\} = 0$. Nevertheless, in view of the celebrated Schauder's Fixed Point Theorem, which states that each continuous self-mapping T defined on compact and convex set has a fixed point, the above result is very weak.

A situation is different in the case of weakly* compact and convex subsets of ℓ_1 considered as a dual of c. To see it, we shall need some basic facts.

Recall that classical Schauder basis of c consists of vectors

$$e = (1, 1, \ldots, 1, \ldots) \text{ and } e^i = (\underbrace{0, \ldots, 0}_{i-1}, 1, 0, \ldots, 0, \ldots) \text{ for } i = 1, 2, \ldots.$$

Then, for each $x = (x_1, x_2, \ldots) \in c$, we have

$$x = \lim_{n \to \infty} x_n \cdot e + \left(x_1 - \lim_{n \to \infty} x_n, x_2 - \lim_{n \to \infty} x_n, \ldots\right)$$

$$= \lim_{n \to \infty} x_n \cdot e + \sum_{i=1}^{\infty} \left(x_i - \lim_{n \to \infty} x_n\right) e^i.$$

Consequently, if $x^* \in c^*$ and $x = (x_1, x_2, \ldots) \in c$, then

$$x^*(x) = \lim_{n \to \infty} x_n \cdot x^*(e) + \sum_{i=1}^{\infty} \left(x_i - \lim_{n \to \infty} x_n\right) x^*(e^i)$$

$$= \lim_{n \to \infty} x_n \cdot \left(x^*(e) - \sum_{i=1}^{\infty} x^*(e^i)\right) + \sum_{i=1}^{\infty} x_i x^*(e^i);$$

observe that $(x^*(e^1), x^*(e^2), \ldots) \in \ell_1$ because $x^* \in c^*$ implies that the restriction of x^* to c_0 is in c_0^*.

Define the mapping $\varphi : c^* \to \ell_1$ by

$$\varphi(x^*) = \left(x^*(e) - \sum_{i=1}^{\infty} x^*(e^i), x^*(e^1), x^*(e^2), \ldots\right).$$

Note that $\varphi(c^*) = \ell_1$. Indeed, if $y = (y_1, y_2, \ldots) \in \ell_1$, then $\varphi(x^*) = y$ for $x^* \in c^*$ such that $x^*(e) = \sum_{i=1}^{\infty} y_i$ and $x^*(e^i) = y_{i+1}$ for $i = 1, 2, \ldots$. We shall prove that φ is an isometry of c^* onto ℓ_1. For $x^* \in c^*$ and $x = (x_1, x_2, \ldots) \in c$, we have

$$|x^*(x)| \le \left|\lim_{n \to \infty} x_n\right| \cdot \left|x^*(e) - \sum_{i=1}^{\infty} x^*(e^i)\right| + \sum_{i=1}^{\infty} |x_i| \cdot |x^*(e^i)|$$

$$\le \left(\left|x^*(e) - \sum_{i=1}^{\infty} x^*(e^i)\right| + \sum_{i=1}^{\infty} |x^*(e^i)|\right) \|x\|.$$

Hence, $\|x^*\|_* \leq \|\varphi(x^*)\|$. On the other hand, if for each $k = 1, 2, \ldots$ we define $x^k = (x_1^k, x_2^k, \ldots) \in c$ as

$$x_i^k = \begin{cases} \operatorname{sgn}(x^*(e^i)), & \text{for } i = 1, \ldots, k \\ \operatorname{sgn}(x^*(e) - \sum_{i=1}^{\infty} x^*(e^i)), & \text{for } i = k+1, k+2, \ldots, \end{cases}$$

then $\|x^k\| = 1$, provided $x^* \neq 0$, and

$$x^*(x^k) = \left| x^*(e) - \sum_{i=1}^{\infty} x^*(e^i) \right| + \sum_{i=1}^{k} |x^*(e^i)| \pm \sum_{i=k+1}^{\infty} x^*(e^i).$$

Thus,

$$\lim_{k \to \infty} x^*(x^k) = \left| x^*(e) - \sum_{i=1}^{\infty} x^*(c^i) \right| + \sum_{i=1}^{\infty} |x^*(e^i)|,$$

and, consequently, $\|x^*\|_* \geq \|\varphi(x^*)\|$. Finally, $\|x^*\|_* = \|\varphi(x^*)\|$. We shall summarize the above remarks in the following

Lemma 7.3 *For each $x^* \in c^*$, there exists exactly one vector*

$$y = (y_1, y_2, \ldots) = \left(x^*(e) - \sum_{i=1}^{\infty} x^*(e^i), x^*(e^1), x^*(e^2), \ldots \right) \in \ell_1$$

such that for any $x = (x_1, x_2, \ldots) \in c$

$$x^*(x) = y_1 \lim_{n \to \infty} x_n + \sum_{n=1}^{\infty} x_n y_{n+1} \quad and \quad \|x^*\|_* = \|y\| = \sum_{n=1}^{\infty} |y_n|.$$

The mapping $\varphi : x^* \mapsto y$ is a linear isometry of c^* onto ℓ_1. Hence, up to isometry, $c^* = \ell_1$.

Immediately, we can ask about the characterization of weakly* convergent sequences in $\ell_1 = c^*$. Suppose $\{y^k\}_{k=1}^{\infty}$ is a bounded sequence in ℓ_1 and let $\{x_k^*\}_{k=1}^{\infty}$ be a sequence in c^* such that $\varphi(x_k^*) = y^k = (y_1^k, y_2^k, \ldots)$. Since vectors

$$e = (1, 1, \ldots, 1, \ldots) \quad \text{and} \quad e^i = (\underbrace{0, \ldots, 0}_{i-1}, 1, 0, \ldots, 0, \ldots) \text{ for } i = 1, 2, \ldots$$

form a linearly dense set in c, the weak* convergence of the sequence $\{y^k\}_{k=1}^{\infty}$ to $y = (y_1, y_2, \ldots) = \varphi(x^*)$ is equivalent to the condition $\lim_{k \to \infty} x_k^*(e) = x^*(e)$ and $\lim_{k \to \infty} x_k^*(e^i) = x^*(e^i)$ for each $i = 1, 2, \ldots$. Since

$$x^*(e^i) = y_{i+1}, \quad x^*(e) = \sum_{n=1}^{\infty} y_n, \quad x_k^*(e) = \sum_{n=1}^{\infty} y_n^k, \quad \text{and} \quad x_k^*(e^i) = y_{i+1}^k$$

for $i, k = 1, 2, \ldots$, we obtain the following:

Lemma 7.4 *Let* $\{y^k\}_{k=1}^{\infty}$, $y^k = (y_1^k, y_2^k, \dots)$, *be a bounded sequence in* ℓ_1. *Then* $\{y^k\}_{k=1}^{\infty}$ *is weakly* convergent to* $y = (y_1, y_2, \dots)$ *in* $\ell_1 = c^*$ *if and only if for each* $i = 1, 2, \dots$

$$\lim_{k \to \infty} y_{i+1}^k = y_{i+1}$$

and

$$\lim_{k \to \infty} \left(\sum_{n=1}^{\infty} y_n^k \right) = \sum_{n=1}^{\infty} y_n.$$

Example 7.3 *Consider the sequence* $\{e^n\}$ *of basis vectors in* ℓ_1. *Then*

$$w^* - \lim_{n \to \infty} e^n = (0, 0, \dots, 0, \dots)$$

in $\ell_1 = c_0^*$. *However, if we consider* ℓ_1 *as a dual of* c, *then*

$$w^* - \lim_{n \to \infty} e^n = e^1 = (1, 0, \dots, 0, \dots).$$

This shows that both weak* topologies are not comparable.

In opposite to $\ell_1 = c_0^*$, the space $\ell_1 = c^*$ does not have the weak*-f.p.p. for nonexpansive mappings. To see it, consider the following:

Example 7.4 *Let* $j : c \to c^{**}$ *be the canonical embedding; that is,* $j(x)(x^*) = x^*(x)$ *for* $x \in c$ *and* $x^* \in c^*$. *Let* $e = (1, 1, \dots, 1, \dots) \in c$, *and consider the set*

$$(j(e))^{-1}\{1\} = \{x^* \in c^* : (j(e))(x^*) = 1\} = \{x^* \in c^* : x^*(e) = 1\}.$$

If φ *denotes an isometry from lemma 7.3, then*

$$\varphi\left((j(e))^{-1}\{1\}\right) = \left\{(x_1, x_2, \dots) \in \ell_1 : x_1 = 1 - \sum_{i=2}^{\infty} x_i\right\}.$$

If B^* *denotes the closed unit ball in* c^*, *then the set*

$$\varphi\left((j(e))^{-1}\{1\}\right) \cap \varphi(B^*)$$

is weakly* compact as the intersection of weakly* closed and weakly* compact sets. It is easy to see that the above set is a "positive face" S^+ of the closed unit ball B in ℓ_1; that is,

$$S^+ = \left\{x = (x_1, x_2, \dots) \in \ell_1 : x_i \geq 0 \text{ for } i = 1, 2, \dots, \text{ and } \sum_{i=1}^{\infty} x_i = 1\right\}.$$

To prove that S^+ fails to have the fixed point property, consider the "right shift" mapping $T : S^+ \to S^+$ defined for any $x = (x_1, x_2, \dots) \in S^+$ by

$$T(x_1, x_2, \dots) = (0, x_1, x_2, \dots).$$

It is easy to verify that T is an isometry; hence, it is nonexpansive. Further, $Tx = x$ implies $x = (0, 0, \ldots, 0, \ldots)$, but $(0, 0, \ldots, 0, \ldots) \notin S^{+}$. Hence, T is fixed point free.

Obviously, S^{+} is not weakly* compact in the weak* topology generated by c_0.

7.2 Minimal invariant sets and normal structure

Another approach was presented by Kirk in 1965. The first main ingredient in his proof is a geometric property called **normal structure**, which was introduced by Brodskii and Milman [15] in 1948.

Let C be a bounded, closed and convex subset of a Banach space $(X, \|\cdot\|)$. For any $x \in C$, we define the *radius of C with respect to x* as the number

$$r_x(C) = \sup \{\|x - y\| : y \in C\}.$$

We say that a point $x \in C$ is a *diametral point* if

$$r_x(C) = \operatorname{diam}(C)$$

and a *non-diametral point* if

$$r_x(C) < \operatorname{diam}(C).$$

If each point of C is its diametral point, then we say that C is a *diametral set*. The *Chebyshev radius of C* (related to C) is the number

$$r(C) = \inf \{r_x(C) : x \in C\},$$

and the *Chebyshev center of C* (related to C) is given by

$$K(C) = \{x \in C : r_x(C) = r(C)\}.$$

The reader can easily verify that $K(C)$ can be given via relation

$$K(C) = \bigcap_{\epsilon > 0} K_\epsilon(C),$$

where

$$K_\epsilon(C) = \{y \in C : r_y(C) \le r(C) + \epsilon\} = \bigcap_{x \in C} B(x, r(C) + \epsilon) \cap C.$$

It may happen that $K(C) = \emptyset$. However, if C is weakly compact, then all of its closed and convex subsets are weakly compact. Hence, $K(C) \ne \emptyset$. If C is weakly* compact subset of X^*, then by the Alaoglu Theorem, all closed balls and, consequently, all sets of the form $B(x, r(C) + \epsilon) \cap C$ are weakly* compact; hence, $K(C) \neq \emptyset$.

Clearly, the simplest diametral sets are singletons. However, there exist bounded, closed and convex sets with a positive diameter, which are diametral!

Example 7.5 *Consider the space c_0 with its usual sup norm,*

$$\|(x_1, x_2, \dots)\| = \max\{|x_i| : i = 1, 2, \dots\}.$$

Let B^+ denote the set

$$B^+ = \left\{ \{x_i\} : x_i \geq 0 \text{ and } \sum_{i=1}^{\infty} x_i \leq 1 \right\}.$$

The reader can easily verify that B^+ is a bounded, closed and convex subset of c_0 with $\mathrm{diam}(B^+) = 1$. We claim that B^+ is diametral. Indeed, if e_n denotes n-th vector of the standard basis, then, for each $x \in B^+$, we have

$$\lim_{n \to \infty} \|x - e_n\| = 1,$$

and, consequently, $r(B^+) = \mathrm{diam}(B^+) = 1$.

Example 7.6 *In the space ℓ_1 furnished with the standard norm,*

$$\|(x_1, x_2, \dots)\| = \sum_{i=1}^{\infty} |x_i|,$$

consider the "positive face" S^+ of the unit sphere S,

$$S^+ = \left\{ \{x_i\} : x_i \geq 0 \text{ and } \sum_{i=1}^{\infty} x_i = 1 \right\}.$$

We leave to the reader to verify that S^+ is a bounded, closed and convex subset of ℓ_1, having diameter equal to 2. Observe that, for each $x \in S^+$,

$$\lim_{n \to \infty} \|x - e_n\| = 2,$$

where e_n denotes n-th vector of the standard basis. Hence, S^+ is a diametral set.

One can observe that, in examples 7.5 and 7.6, we have

$$B^+ = \overline{\mathrm{conv}}\{e_n : n = 1, 2, \dots\}$$

and

$$S^+ = \overline{\mathrm{conv}}\{e_n : n = 1, 2, \dots\}.$$

Moreover, in both cases, we have

$$\lim_{n \to \infty} \mathrm{dist}(e_{n+1}, \mathrm{conv}\{e_1, e_2, \dots, e_n\}) = \mathrm{diam}(\{e_1, e_2, \dots\}).$$

On the other hand, we have the following:

Example 7.7 *Consider the space l_2 with the standard norm,*

$$\|x\| = \|(x_1, x_2, \dots)\| = \left(\sum_{i=1}^{\infty} x_i^2 \right)^{\frac{1}{2}}.$$

Let

$$S^+ = \left\{ \{x_i\} : x_i \geq 0, \sum_{i=1}^{\infty} x_i \leq 1 \right\}.$$

One can notice that S^+ is bounded, closed and convex in l_2. Moreover, $\mathrm{diam}\,(S^+) = \sqrt{2}$, whereas $r\,(S^+) = 1$. Hence, S^+ is not diametral. Also,

$$S^+ = \overline{conv}\,\{e_n : n = 1, 2, \dots\},$$

where e_n denotes n-th vector of the standard Schauder basis in l_2. Nevertheless,

$$\lim_{n \to \infty} dist\,(e_{n+1}, conv\,\{e_1, e_2, \dots, e_n\}) = r(S^+) = 1 < diam\,(\{e_1, e_2, \dots\}) = \sqrt{2}.$$

We shall see below that it is not a coincidence. To proceed, we shall need some definitions introduced by Brodskii and Milman [15].

We say that a closed and convex set $D \subset X$ (or the space X itself) has *normal structure* if any bounded, closed and convex subset C of D with $\mathrm{diam}(C) > 0$ contains a non-diametral point.

A bounded sequence $\{x_n\}$ in a Banach space X is said to be *a diametral sequence* if it is not constant and

$$\lim_{n \to \infty} dist\,(x_{n+1}, conv\,\{x_1, x_2, \dots, x_n\}) = diam\,(\{x_1, x_2, \dots\}).$$

Lemma 7.5 (Brodskii and Milman, 1948) *Let C be a bounded, closed and convex subset of a Banach space X. Then the following are equivalent.*

(i) C has normal structure.

(ii) C does not contain a diametral sequence.

Proof. First, we shall prove that not (ii) implies not (i). Indeed, if C contains a diametral sequence $\{x_n\}$, then the set $D = \overline{conv}\,\{x_1, x_2, \dots\}$ is a diametral subset of C. Hence, C does not have normal structure.

To prove that not (i) implies not (ii), suppose that D is a diametral subset of C. Let $d = \mathrm{diam}(D) > 0$ and $\epsilon \in (0, d)$. Let $y_1 := x_1 \in D$. Since D is diametral, there is a point x_2 such that $\|x_2 - y_1\| > d - \epsilon$. Define y_2 as

$$y_2 = \frac{x_1 + x_2}{2}.$$

Further, since D is diametral, there exists a point $x_3 \in D$ such that

$\|x_3 - y_2\| > d - \frac{\epsilon}{4}$. Inductively, we construct a sequence $\{x_n\}$ of elements of D satisfying

$$\|x_{n+1} - y_n\| > d - \frac{\epsilon}{n^2},$$

where

$$y_n = \frac{x_1 + x_2 + \cdots + x_n}{n}.$$

We shall prove that $\{x_n\}$ is diametral. Take any $x \in \text{conv}\{x_1, \ldots, x_n\}$. Then $x = \sum_{i=1}^{n} \alpha_i x_i$ for some $\alpha_i \geq 0$ such that $\sum_{i=1}^{n} \alpha_i = 1$. Define $\widehat{\alpha} := \max\{\alpha_1, \ldots, \alpha_n\}$. It is clear that $\widehat{\alpha} > 0$. Now, we can express y_n as

$$y_n = \frac{1}{n\widehat{\alpha}} x + \sum_{i=1}^{n} \left(\frac{1}{n} - \frac{\alpha_i}{n\widehat{\alpha}} \right) x_i$$

and observe that

$$\frac{1}{n} - \frac{\alpha_i}{n\widehat{\alpha}} \geq 0 \quad \text{and} \quad \frac{1}{n\widehat{\alpha}} + \sum_{i=1}^{n} \left(\frac{1}{n} - \frac{\alpha_i}{n\widehat{\alpha}} \right) = 1.$$

We have

$$d - \frac{\epsilon}{n^2} < \|y_n - x_{n+1}\|$$

$$\leq \frac{1}{n\widehat{\alpha}} \|x - x_{n+1}\| + \sum_{i=1}^{n} \left(\frac{1}{n} - \frac{\alpha_i}{n\widehat{\alpha}} \right) \|x_i - x_{n+1}\|$$

$$\leq \frac{1}{n\widehat{\alpha}} \|x - x_{n+1}\| + \left(1 - \frac{1}{n\widehat{\alpha}} \right) d.$$

Thus,

$$\|x - x_{n+1}\| \geq \left(\frac{d}{n\widehat{\alpha}} - \frac{\epsilon}{n^2} \right) n\widehat{\alpha}$$

$$= d - \frac{\epsilon\widehat{\alpha}}{n}$$

$$\geq d - \frac{\epsilon}{n}.$$

Since $x \in \text{conv}\{x_1, \ldots, x_n\}$ and $\epsilon \in (0, d)$ has been arbitrarily chosen, we conclude that for any $n \in \mathbb{N}$,

$$d \geq \text{dist}\,(x_{n+1}, \text{conv}\{x_1, x_2, \ldots, x_n\}) \geq d - \frac{\epsilon}{n}.$$

Therefore, $\{x_n\}$ is a diametral sequence.

$\square$

One can observe that a diametral sequence does not contain a convergent subsequence. This implies that all compact and convex subsets of X have normal structure. In particular, every finite dimensional Banach space X has normal structure. Furthermore, we have the following:

Lemma 7.6 *If X is a Banach space with a characteristic of convexity $\epsilon_0(X) < 1$, then X has normal structure. In particular, all uniformly convex Banach spaces have normal structure.*

Proof. Let $C \subset X$ be a bounded, closed and convex set with a positive diameter $d = \operatorname{diam}(C)$. Fix $\epsilon \in (0, d\,(1 - \epsilon_0(X)))$ and two points u and v in C such that $\|u - v\| > d - \epsilon$. The following implication follows from (2.2) and the fact that the modulus of convexity δ_X is nondecreasing on $[0, 2]$: for any $x \in C$,

$$\left.\begin{array}{c} \|x - u\| \leq d \\ \|x - v\| \leq d \end{array}\right\} \implies \left\| x - \frac{u + v}{2} \right\| \leq \left(1 - \delta_X\left(\frac{d - \epsilon}{d}\right)\right) d. \qquad (7.4)$$

Since $\frac{d-\epsilon}{d} > \epsilon_0(X)$ we get $\delta_X\left(\frac{d-\epsilon}{d}\right) > 0$; hence, $\frac{u+v}{2}$ is a non-diametral point of C.

$\square$

The second main ingredient in Kirk's proof is Zorn's Lemma, which ensures the existence of the so-called **minimal invariant set**. Let C be a weakly compact and convex subset of a Banach space X and $T : C \to C$ be a mapping. A nonempty, weakly closed and convex subset C° of C is called *minimal invariant* for T if $T(C^\circ) \subset C^\circ$ and C° does not contain a nonempty, weakly closed and convex proper subset D, which is T-*invariant*; that is, $T(D) \subset D$.

Recall that by a relation $\preceq$ on a set A we mean a subset of $A \times A$. For $a, b \in A$, we write $a \preceq b$ if and only if $(a, b) \in\ \preceq$. We say that $(A, \preceq)$ is a *partially ordered set* if for each $a, b, c \in A$:

(i) $a \preceq a$,

(ii) $(a \preceq b$ and $b \preceq a) \Rightarrow a = b$,

(iii) $(a \preceq b$ and $b \preceq c) \Rightarrow a \preceq c$.

Any subset $B \subset A$, which is *linearly ordered* by $\preceq$ is called a *chain*; recall that $(B, \preceq)$ is linearly ordered if it is partially ordered, and, in addition, for each $a, b \in B$:

(iv) $a \preceq b$ or $b \preceq a$.

Let $(A, \preceq)$ be a partially ordered set. We say that $a \in A$ is *maximal* if

$$\forall_{b \in A}\,(a \preceq b \Rightarrow a = b).$$

Let $B \subset A$. An element $a \in A$ is called an *upper bound* for B if

$$\forall_{b \in B}\,(b \preceq a).$$

Lemma 7.7 (Zorn's Lemma) *Let $(A, \preceq)$ be a partially ordered set. If every chain has an upper bound, then A has a maximal element.*

Now we are ready to present Kirk's Theorem.

Theorem 7.3 (Kirk's Theorem, 1965) *Let C be a nonempty, weakly compact and convex subset of a Banach space X and suppose that C has normal structure. Then each nonexpansive mapping $T : C \to C$ has a fixed point.*

Proof. Let $\mathfrak{F}$ denote the family of all nonempty, weakly closed and convex subsets $D \subset C$ such that $T(D) \subset D$. It is clear that the relation on $\mathfrak{F}$ defined as

$$D_1 \preceq D_2 \Leftrightarrow D_1 \supset D_2$$

generates a partial order. Consider any chain $\mathcal{D} \subset \mathfrak{F}$ and the set

$$D^\circ = \bigcap_{D \in \mathcal{D}} D.$$

Since C is weakly compact and the family $\mathcal{D}$ has the finite intersection property, we conclude that $D^\circ \neq \emptyset$. Moreover, D° is weakly closed, convex and $T(D^\circ) \subset D^\circ$. Hence, D° is an upper bound for the chain $\mathcal{D}$. By Zorn's Lemma, there is a maximal element $C^\circ \in \mathfrak{F}$. Put

$$C^{\circ\circ} = \overline{\mathrm{conv}}T(C^\circ) \subset C^\circ.$$

Then,

$$T(C^{\circ\circ}) \subset T(C^\circ) \subset \overline{\mathrm{conv}}T(C^\circ) = C^{\circ\circ}.$$

Consequently, $C^{\circ\circ} \in \mathfrak{F}$. By maximality of C°, it must be the case $C^\circ = C^{\circ\circ}$. Thus,

$$\overline{\mathrm{conv}}T(C^\circ) = C^\circ.$$

Since C° is weakly compact, we conclude that

$$K(C^\circ) = \{z \in C^\circ : r_z(C^\circ) = r(C^\circ)\} \neq \emptyset;$$

hence, there exists a point $z \in C^\circ$ such that $r_z(C^\circ) = r(C^\circ)$. Observe that, for each $y \in C^\circ$, we have

$$\|Tz - Ty\| \leq \|z - y\| \leq r_z(C^\circ) = r(C^\circ)$$

for all $y \in C^\circ$. Hence,

$$T(C^\circ) \subset B(T(z), r(C^\circ)),$$

and, consequently,

$$C^\circ = \overline{\mathrm{conv}}T(C^\circ) \subset B(T(z), r(C^\circ)).$$

Thus,
$$r_{Tz}(C^\circ) = r_z(C^\circ) = r(C^\circ).$$
This implies that $K(C^\circ)$ is T-invariant. The set C° is minimal invariant; hence,
$$K(C^\circ) = C^\circ.$$
Since C° has normal structure, we conclude that $\operatorname{diam}(C^\circ) = 0$. Thus,
$$C^\circ = \{z\},$$
and z is a fixed point of T.

$\square$

If X is reflexive Banach space, then any bounded, closed and convex subset C of X is weakly compact. Hence, we obtain the following:

Corollary 7.1 *Let X be reflexive Banach space. If X has normal structure, then X has the f.p.p. for nonexpansive mappings.*

Example 7.8 *Consider the space c_0 renormed as in example 2.2; that is,*

$$|||x||| = \left(\|x\|^2 + \sum_{i=1}^{\infty} \left(\frac{x_i}{i} \right)^2 \right)^{\frac{1}{2}}.$$

Then, $(c_0, ||| \cdot |||)$ has normal structure. To see it, consider any bounded, closed and convex set $C \subset c_0$ with $\operatorname{diam}(C) > 0$. Fix any two points $u = (u_1, u_2, \dots)$ and $v = (v_1, v_2, \dots)$ in C such that $u \neq v$. Let $z = \frac{u+v}{2}$. We shall prove that z is a non-diametral point of C. For any $x \in C$, we have

$$|||x - z|||^2 = \|x - z\|^2 + \sum_{i=1}^{\infty} \left(\frac{x_i - z_i}{i} \right)^2.$$

Observe that

$$
\begin{aligned}
\|x - z\|^2 &= \frac{1}{4} \|(x - u) + (x - v)\|^2 \\
&\leq \frac{1}{4} (\|x - u\| + \|x - v\|)^2 \\
&= \frac{1}{2} \left(\|x - u\|^2 + \|x - v\|^2 \right) - \frac{1}{4} (\|x - u\| - \|x - v\|)^2 \\
&\leq \frac{1}{2} \left(\|x - u\|^2 + \|x - v\|^2 \right)
\end{aligned}
$$

and for each $i = 1, 2, \dots,$

$$
\begin{aligned}
\left(\frac{x_i - z_i}{i} \right)^2 &= \frac{1}{4} \left(\frac{(x_i - u_i) + (x_i - v_i)}{i} \right)^2 \\
&= \frac{1}{2} \left[\left(\frac{x_i - u_i}{i} \right)^2 + \left(\frac{x_i - v_i}{i} \right)^2 \right] - \frac{1}{4} \left(\frac{u_i - v_i}{i} \right)^2.
\end{aligned}
$$

Hence,

$$|||x-z|||^2 \leq \frac{1}{2}\left(|||x-u|||^2 + |||x-v|||^2\right) - \frac{1}{4}\sum_{i=1}^{\infty}\left(\frac{u_i-v_i}{i}\right)^2$$

$$\leq d^2 - \frac{1}{4}\sum_{i=1}^{\infty}\left(\frac{u_i-v_i}{i}\right)^2.$$

Since $u \neq v$ are fixed and $x \in C$ has been arbitrary chosen, we conclude that $r(C) \leq r_z(C) < diam(C)$.

In particular, the set

$$B^+ = \{x = \{x_i\} : \|x\| \leq 1, \ x_i \geq 0 \ for \ i = 1, 2, \dots\}$$

has normal structure with respect to the new norm $||| \cdot |||$, and it is easy to observe that the mapping $T : B^+ \to B^+$ defined as

$$T(x_1, x_2, \dots) = (1, x_1, x_2, \dots)$$

is nonexpansive with respect to $||| \cdot |||$, and it is fixed point free. Hence, T does not have a minimal invariant set. Consequently, normal structure of the set or the space without some additional assumption does not imply the fixed point property.

Also, the above example shows that normal structure of X does not imply reflexivity. Moreover, Zizler [77] proved that any separable Banach space can be renormed to have normal structure. Actually, these two properties, normal structure and reflexivity, are independent.

Example 7.9 *(see [31]) In the space ℓ_2 with the classical norm*

$$\|x\| = \|(x_1, x_2, \dots)\| = \left(\sum_{i=1}^{\infty} x_i^2\right)^{\frac{1}{2}},$$

consider the positive part of the unit ball,

$$B^+ = \{x \in \ell_2 : \|x\| \leq 1, \ x_i \geq 0\}.$$

Define a new norm by

$$|||x||| = \max\left\{\frac{1}{\sqrt{2}}\|x\|, \max_{i \in \mathbb{N}}|x_i|\right\}.$$

Observe that $||| \cdot |||$ is equivalent to $\|\cdot\|$ and for any $x \in \ell_2$

$$\frac{1}{\sqrt{2}}\|x\| \leq |||x||| \leq \|x\|.$$

Hence, $(\ell_2, ||| \cdot |||)$ is reflexive. However, it does not have normal structure. We leave to the reader to verify that the set B^+ is diametral with respect to the new norm $||| \cdot |||$ and $r(B^+) = diam(B^+) = 1$.

In fact, the estimate (7.4) does not depend on the choice of a set C. This leads us to the following definition introduced in 1980 by Bynum [18].

We say that a closed, convex set $D \subset X$ (or the whole space X) has *uniform normal structure* if there is a constant $a \in (0,1)$ such that, for any bounded, closed and convex subset C of D, we have

$$r(C) \leq a \cdot \operatorname{diam}(C).$$

The number

$$N(X) := \inf \left\{ \frac{\operatorname{diam}(C)}{r(C)} : C \subset X \text{ is bounded, closed and convex} \right.$$
$$\left. \text{with } \operatorname{diam}(C) > 0 \right\}$$

is called the *normal structure coefficient* of the space X. Equivalently, a Banach space $(X, \|\cdot\|)$ has uniform normal structure if and only if $N(X) > 1$. From (7.4), we conclude that, if $\epsilon_0(X) < 1$, then X has uniform normal structure with $a = 1 - \delta_X(1)$, and, consequently,

$$N(X) \geq \frac{1}{1 - \delta_X(1)}.$$

Moreover, Bynum [18] proved that for a Hilbert space H we have $N(H) = \sqrt{2}$. Domínguez Benavides [7] and Prus [71] gave exact values of the constant $N(X)$ in the case of ℓ_p and $L_p[0,1]$ spaces:

$$N(\ell_p) = N(L_p[0,1]) = \min \left\{ 2^{\frac{1}{p}}, 2^{\frac{1}{q}} \right\},$$

where $p > 1$ and $\frac{1}{p} + \frac{1}{q} = 1$. It is clear that uniform normal structure implies normal structure. In fact, it is a much stronger property. In 1984, Maluta [58] proved that every Banach space X having uniform normal structure must be reflexive!

Corollary 7.2 *If X has uniform normal structure, then X has the f.p.p. for nonexpansive mappings.*

Let us pass now to the case of weakly* compact sets in dual space X^*. Let D be a convex subset of a dual Banach space X^*. We say that D (or the space X^* itself) has *weak* normal structure* if each bounded, weakly* closed and convex set $E \subset D$ with $\operatorname{diam}(E) > 0$ contains a non-diametral point.

Theorem 7.4 (Kirk's Theorem) *Let X be a Banach space. Let C be a weakly* compact convex subset of a dual Banach space X^* and suppose C has weak* normal structure. Then, C has f.p.p. for nonexpansive mappings.*

Proof. Repeat steps from the proof of theorem 7.3.

□

Now it is enough to apply the following lemma to obtain another proof of Theorem 7.2:

Lemma 7.8 *If C is a weakly* compact and convex subset of $l_1 = c_0^*$, then C has weak* normal structure.*

Proof. Obviously, it is enough to prove that every weakly* compact and convex set C in $l_1 = c_0^*$ contains a non-diametral point. Suppose $\text{diam}(C) > 0$. If C is compact, then it has weak* normal structure. Hence, suppose C is not compact. Then, there exists a number $a > 0$ and a sequence $\{x^n\}$ in C such that $\|x^i - x^j\| \geq a$ for all $i, j = 1, 2, \ldots$ with $i \neq j$. Since the weak* topology on C is metrizable, we conclude that $\{x^n\}$ contains a weakly* convergent subsequence to $x \in C$. We can assume that $\{x^n\}$ has been already chosen to satisfy this condition. It is clear that $\{x^n\}$ does not converge in norm to x. This implies that

$$r\left(x, \{x^n\}\right) = \limsup_{n \longrightarrow \infty} \|x^n - x\| > 0.$$

By lemma 7.2, for each $y \in C$, we have

$$r\left(y, \{x^n\}\right) = r\left(x, \{x^n\}\right) + \|x - y\|,$$

and, consequently,

$$\|x - y\| = r\left(y, \{x^n\}\right) - r\left(x, \{x^n\}\right) \leq \text{diam}(C) - r\left(x, \{x^n\}\right).$$

Hence, the point x is a non-diametral point of C.

$\square$

We have already seen in the proof of Kirk's Theorem that any minimal invariant set C must be diametral. However, every such set possesses a much stronger property, which has proved to be very useful in the study of fixed point property for nonexpansive mappings. This fact has been discovered independently by Goebel [30] and Karlovitz [46]:

Lemma 7.9 (Goebel-Karlovitz Lemma) *Let $C \subset X$ be a minimal with respect to being nonempty, weakly compact, convex, and T-invariant for some nonexpansive mapping T. If $\{x_n\} \subset C$ is an approximate fixed point sequence for T, then for each $x \in C$,*

$$r\left(x, \{x_n\}\right) = \limsup_{n \to \infty} \|x - x_n\| = \lim_{n \to \infty} \|x - x_n\| = diam(C).$$

Proof. It is enough to consider the case when $\text{diam}(C) > 0$. The set C must be diametral. Suppose there is a point $z \in C$ such that $\limsup_{n \longrightarrow \infty} \|z - x_n\| = r < \text{diam}(C)$. Then, the set

$$A_r\left(\{x_n\}, C\right) := \{y \in C : r\left(y, \{x_n\}\right) \leq r\}$$

is nonempty, closed and convex. Since $A_r\left(\{x_n\},C\right)$ is a weakly closed subset of a weakly compact set C, we conclude that $A_r\left(\{x_n\},C\right)$ is itself weakly compact. Also, the set $A_r\left(\{x_n\},C\right)$ is T-invariant. Hence, by minimality of C, we have $A_r\left(\{x_n\},C\right)=C$. Further, we can select $\mu>0$ such that $r+\mu<$ diam(C). Choose any finite collection of points $\{y_1,y_2,\ldots,y_m\}\subset C$. Then, there exists an integer k such that for each $n>k$ and for each $i=1,2,\ldots,m$, we have $x_n\in B\left(y_i,r+\mu\right)\cap C$. Thus, the family $\{B(x,r+\mu)\cap C:x\in C\}$ has the finite intersection property. Since all sets of the form $B(x,r+\mu)\cap C$ are weakly compact, we conclude that

$$\bigcap_{x\in C}\left(B(x,r+\mu)\cap C\right)=C\cap\bigcap_{x\in C}B(x,r+\mu)\neq\emptyset.$$

Then, for any $z\in C\cap\bigcap_{x\in C}B(x,r+\mu)$, we have $r(z,C)\leq r+\mu<$ diam(C). However, this contradicts the fact that C is diametral. Hence, for each $x\in C$, we have

$$\limsup_{n\to\infty}\|x-x_n\|=\text{diam}(C).$$

Since each subsequence of $\{x_n\}$ is still an approximate fixed point sequence for T, we can replace lim sup by lim.

$\square$

For several years, it had been unknown whether there exists a weakly compact and convex set C with a positive diameter that is minimal invariant for a certain nonexpansive mapping $T:C\to C$. The answer came in 1981 when Alspach [1] published an example of fixed point free self-mapping defined on a weakly compact and convex set.

Example 7.10 (Alspach's example) *Let $X=L_1[0,1]$. Recall that for any two functions $g,h\in L_1[0,1]$ satisfying $g\leq h$ up to the set of measure zero, the order segment, defined as*

$$\{f\in L_1[0,1]:g(t)\leq f(t)\leq h(t)\text{ almost everywhere in }[0,1]\},$$

is weakly compact in $L_1[0,1]$.
    Let C be the set

$$C=\left\{f\in L_1[0,1]:0\leq f\leq 1\text{ and }\int_0^1 f(t)\,dt=\frac{1}{2}\right\}.$$

The set C is convex and weakly compact as the intersection of the order segment with a hyperplane. Consider the following mapping, usually called the "baker transformation",

$$(Tf)(t)=\begin{cases}\min\{2f(2t),1\} & \text{if }0\leq t\leq\frac{1}{2}\\ \max\{2f(2t-1)-1,0\} & \text{if }\frac{1}{2}<t\leq 1.\end{cases}$$

We leave to the reader the verification that T is an isometry on C. We shall prove that T is fixed point free. Suppose $g \in C$ is a fixed point of T, $g = Tg$. Then,

$$A := \{t : g(t) = 1\} = \{t : (Tg)(t) = 1\},$$

and, by definition of T, we get

$$A = \left\{t : 0 \leq t \leq \frac{1}{2} \wedge \frac{1}{2} \leq g(2t) \leq 1\right\} \cup \left\{t : \frac{1}{2} < t \leq 1 \wedge g(2t-1) = 1\right\}$$

$$= \frac{1}{2}\left\{t : \frac{1}{2} \leq g(t) < 1\right\} \cup \frac{1}{2}\{t : g(t) = 1\} \cup \left\{\frac{1+t}{2} : g(t) = 1\right\}.$$

Observe that the above three sets are mutually disjoint. Moreover, the measure of each of the last two sets is equal to the measure of A. Consequently, the set

$$\left\{t : \frac{1}{2} \leq g(t) < 1\right\}$$

has measure zero. Further,

$$\left\{\frac{t}{2} : \frac{1}{4} \leq g(t) < \frac{1}{2}\right\} \subset \left\{t : \frac{1}{2} \leq Tg(t) < 1\right\} = \left\{t : \frac{1}{2} \leq g(t) < 1\right\}.$$

Hence, the measure of

$$\left\{t : \frac{1}{4} \leq g(t) < \frac{1}{2}\right\}$$

is zero. Continuing the process, it follows that the measure of each set of the form

$$\left\{t : \frac{1}{2^n} \leq g(t) < \frac{1}{2^{n-1}}\right\}$$

equals zero for $n = 1, 2, \ldots$, and, consequently, the same holds for the set

$$\{t : 0 < g(t) < 1\} = \bigcup_{n=1}^{\infty} \left\{t : \frac{1}{2^n} \leq g(t) < \frac{1}{2^{n-1}}\right\}.$$

Hence,

$$g(t) = \chi_A(t) = \begin{cases} 1 & \text{if } t \in A \\ 0 & \text{if } t \in [0,1] \setminus A. \end{cases}$$

Since

$$\int_0^1 \chi_A(t)dt = \frac{1}{2},$$

the measure of A is $\frac{1}{2}$. We have

$$T(\chi_A) = \chi_{\frac{1}{2}A} + \chi_{(\frac{1}{2} + \frac{1}{2}A)} = \chi_A.$$

Hence, the intersections of A with each of the intervals $\left[0, \frac{1}{2}\right]$ and $\left[\frac{1}{2}, 1\right]$ have measure equal to $\frac{1}{2}$. Further,

$$T^2\left(\chi_A\right) = \chi_{\frac{1}{4}A} + \chi_{\left(\frac{1}{4} + \frac{1}{4}A\right)} + \chi_{\left(\frac{1}{2} + \frac{1}{4}A\right)} + \chi_{\left(\frac{3}{4} + \frac{1}{4}A\right)} = \chi_A.$$

This implies that the intersections of A with each of the intervals $\left[0, \frac{1}{4}\right]$, $\left[\frac{1}{4}, \frac{1}{2}\right]$, $\left[\frac{1}{2}, \frac{3}{4}\right]$, $\left[\frac{3}{4}, 1\right]$ have measure equal to one half that of the interval. Continuing in this way we have that the same is true for any dyadic interval $\left[\frac{i}{2^n}, \frac{i+1}{2^n}\right]$, $n = 1, 2, \ldots$, $i = 0, 1, 2, \ldots, 2^n - 1$, and hence for any interval. Thus, for any $t \in (0, 1)$, the density of A at t defined as

$$\lim_{h \to 0^+} \frac{\mu\left(A \cap [t - h, t + h]\right)}{2h}$$

is equal to $\frac{1}{2}$. However, it is impossible because, in view of the Banach's Density Theorem, any measurable set has density 0 or 1 almost everywhere.

One can notice that the Chebyshev radius $r(C) = \frac{1}{2}$, whereas $\text{diam}(C) = 1$. Consequently, C is not diametral; hence, it is not a minimal invariant set for T.

Consequently, weak compactness itself does not imply the f.p.p.

7.3 Uniformly nonsquare, uniformly noncreasy, and reflexive Banach spaces

We have already seen that all Banach spaces with the characteristic of convexity $\epsilon_0(X) < 1$ share the fixed point property for nonexpansive mappings. In 2005, this result was extended for all uniformly nonsquare Banach spaces. In [28], García-Falset, Llorens-Fuster, and Mazcuñán-Navarro proved the following:

Theorem 7.5 *Let $(X, \|\cdot\|)$ be a Banach space with $\epsilon_0(X) < 2$. Then, any bounded, closed and convex subset C has the f.p.p. for nonexpansive mappings.*

Another approach has been presented by Prus [72]. He introduced a class of *uniformly noncreasy* Banach spaces. For any $x^*, y^* \in S_{X^*}$ and $\delta \in [0, 1]$, $S(x^*, y^*, \delta)$ denotes the set

$$S(x^*, \delta) \cap S(y^*, \delta) = \{x \in B_X : x^*(x) \geq 1 - \delta\} \cap \{x \in B_X : y^*(x) \geq 1 - \delta\}.$$

A closed unit ball B_X (or S_X) is said to have a *crease* if, for some $x^*, y^* \in S_{X^*}$, we have $\text{diam}(S(x^*, y^*, 0)) > 0$ provided $x^* \neq y^*$. A Banach space is named uniformly noncreasy if the following implication holds: for any $\epsilon > 0$, there is $\delta > 0$ such that for any $x^*, y^* \in X^*$

$$\left. \begin{array}{l} \|x^*\|_* = 1 \\ \|y^*\|_* = 1 \\ \|x^* - y^*\|_* > \epsilon \end{array} \right\} \implies \mathrm{diam}(S(x^*, y^*, \delta)) \le \epsilon.$$

Prus proved the following:

Theorem 7.6 *If X is uniformly noncreasy Banach space, then X has the f.p.p. for nonexpansive mappings.*

It is known that uniformly nonsquare as well as uniformly noncreasy Banach spaces are super-reflexive, but it is still unknown whether super-reflexivity implies the fixed point property for nonexpansive mappings. However, in 2009, Domíngucz Benavides [9] obtained the following intriguing theorem:

Theorem 7.7 *Let $(X, \|\cdot\|)$ be a reflexive Banach space. Then, there exists an equivalent norm $\|\|\cdot\|\|$ on X such that $(X, \|\|\cdot\|\|)$ has the fixed point property for $\|\|\cdot\|\|$-nonexpansive mappings.*

Proofs of the above-mentioned results go beyond frames of presented monograph, and we do not establish them here.

7.4   Remarks on the stability of f.p.p.

Actually, bounded, closed and convex subsets of a Banach space X with $\epsilon_0(X) < 1$ or $N(X) > 1$ have the f.p.p. not only for nonexpansive mappings but also for uniformly k-lipschitzian mappings, with a constant k slightly greater than 1. The first result of this type was discovered by Goebel and Kirk [35] in 1973:

Theorem 7.8 (Goebel and Kirk) *Let X be a uniformly convex Banach space with the modulus of convexity δ_X, and let C be a bounded, closed and convex subset of X. Suppose that γ satisfies*

$$\gamma\left(1 - \delta_X\left(\frac{1}{\gamma}\right)\right) = 1. \tag{7.5}$$

If $T : C \to C$ is uniformly lipshitzian relative to a constant $k < \gamma$, then T has a fixed point in C.

Recall that, for a Hilbert space H, we have

$$\delta_H(\epsilon) = 1 - \sqrt{1 - \frac{\epsilon^2}{4}}.$$

Solving the equation (7.5) we get

$$\gamma = \frac{\sqrt{5}}{2}.$$

Later, in 1975, the above theorem was improved and generalized by Lifshitz [54]. Let (M, ρ) be a complete metric space. We say that balls in M are c-*regular* if the following holds:

$$\forall_{k \in (0,c)} \exists_{\alpha, \mu \in (0,1)} \forall_{x,y \in M} \forall_{r>0} \rho(x,y) \geq (1 - \mu) r$$
$$\Rightarrow \exists_{z \in M} B\left(x, (1+\mu)r\right) \cap B\left(y, k(1+\mu)r\right) \subset B\left(z, \alpha r\right).$$

The largest number c for which balls in M are c-regular is called *the Lifshitz character of* M and is denoted by $\kappa(M)$; that is,

$$\kappa(M) = \sup\{c > 0: \text{ the balls in } M \text{ are c-regular}\}.$$

It is easy to check that $\kappa(M) \geq 1$ for any M.

Under the above settings, the Lifshitz Theorem states:

Theorem 7.9 (Lifshitz, 1975) *Let (M, ρ) be a bounded and complete metric space. Suppose $T : M \to M$ is uniformly k-lipschitzian mapping with $k < \kappa(M)$. Then, T has a fixed point in M.*

Looking up the proof of this theorem, it can be stated:

Theorem 7.10 (Lifshitz, strong version) *Let (M, ρ) be a bounded and complete metric space. Suppose $T : M \to M$ is uniformly lipschitzian with $k_\infty(T) < \kappa(M)$. Then, T has a fixed point in M.*

Proof. If $\kappa(M) = 1$, then, in view of theorem 6.4, $k_0(T) < 1$. Thus, using theorem 6.2, T has a unique fixed point.

Now, suppose $\kappa(M) > 1$. For fixed $x \in M$, set

$$r(x) = \inf\{r > 0: \text{ for some } y \in M, \rho(x, T^n y) \leq r \text{ for } n = 1, 2, \dots\}.$$

Fix $k \in (k_\infty(T), \kappa(M))$ and let $n_0 \in \mathbb{N}$ be such that, for each $n > n_0$, we have $k(T^n) \leq k$. Let μ and $\alpha \in (0,1)$ be associated with k as in the definition of $\kappa(M)$-regular balls. Then, by definition of $r(x)$, there exists a natural number $m > n_0$ that satisfies

$$\rho(x, T^m x) \geq (1 - \mu)r(x),$$

and there is $y \in M$ such that, for $n = 1, 2, \dots$,

$$\rho(x, T^n y) \leq (1 + \mu)r(x).$$

Thus, by definition of $\kappa(M)$-regular balls, there exists $z \in M$ such that

$$B\left(x,(1+\mu)\,r(x)\right) \cap B\left(T^m x, k\,(1+\mu)\,r(x)\right) \subset B\left(z, \alpha r(x)\right).$$

Since for $n > m$,

$$\rho(T^m x, T^n y) \leq k\rho(x, T^{n-m}y) \leq k(1+\mu)r(x),$$

the orbit $\{T^n y : n > m\}$ is contained in

$$B\left(x,(1+\mu)\,r(x)\right) \cap B\left(T^m x, k\,(1+\mu)\,r(x)\right),$$

and, consequently, in $B\left(z, \alpha r(x)\right)$. Thus,

$$r(z) \leq \alpha r(x).$$

Moreover, for any $u \in B\left(x,(1+\mu)\,r(x)\right) \cap B\left(T^m x, k\,(1+\mu)\,r(x)\right)$,

$$\rho(z,x) \leq \rho(z,u) + \rho(u,x) \leq \alpha r(x) + (1+\mu)r(x) \leq 3r(x).$$

By putting $x_0 = x$, we can construct a sequence $\{x_n\}_{n=0}^{\infty}$, with $x_{n+1} = z(x_n)$, where $z(x_n)$ is obtain via the above procedure. Thus,

$$r(x_n) \leq \alpha^n r(x_0)$$

and

$$\rho(x_{n+1}, x_n) \leq 3r(x_n) \leq 3\alpha^n r(x_0).$$

This implies that $\{x_n\}_{n=0}^{\infty}$ is a Cauchy sequence. Since M is complete, $\{x_n\}_{n=0}^{\infty}$ converges. It is clear that its limit is a fixed point of T.

$\square$

Let us return to our standard situation. Let C be a bounded, closed and convex subset of Banach space X. It is easy to verify that, under this setting, the Lifshitz character $\kappa(X)$ can be defined as

$$\kappa(X) = \sup\left\{c > 0 : r(B(0,1) \cap B(x,c)) < 1 \text{ for all } x \text{ with } \|x\| = 1\right\}.$$

By definition, we conclude that $1 \leq \kappa(X) \leq 2$. Now the Lifshitz Theorem takes the form:

Theorem 7.11 (Lifshitz, strong version) *If $T : C \to C$ is uniformly lipschitzian with $k_{\infty}(T) < \kappa(X)$, then T has a fixed point in C.*

Proof. It is enough to repeat steps from the proof of theorem 7.10 with the point z, which lies in the segment $[x, T^m x]$. Since C is convex, $z \in C$.

$\square$

It is known that, in the case of Hilbert space, we have

$$\kappa(H) = \sqrt{2} > \frac{\sqrt{5}}{2}.$$

In [25], Downing and Turett gave a characterization of spaces with $\kappa(X) > 1$, using the coefficient $\epsilon_0(X)$:

Theorem 7.12 (Downing, Turett) *In any Banach space X,*

$$\kappa(X) > 1 \iff \epsilon_0(X) < 1.$$

In principle, $\kappa(X) > 1$ for every uniformly convex space X ($\epsilon_0(X) = 0$).
 It is natural to define the following constant: for a Banach space X,

$$\gamma_0(X) := \sup \{\, k \,:\, \text{any bounded, closed and convex set } C \subset X$$
$$\text{has the f.p.p. for uniformly } k\text{-lipschitzian mappings} \}.$$

Hence, for any Banach space X with $\epsilon_0(X) < 1$, we have $\gamma_0(X) > 1$, and for a Hilbert space H, $\gamma_0(H) \geq \sqrt{2}$.
 Recall that every uniformly k-lipschitzian mapping $T : C \to C$ generates the metric d on C defined for any $x, y \in C$ by

$$d(x, y) = \sup \{\, \|T^i x - T^i y\| : i = 0, 1, 2, \dots \},$$

which is equivalent to the metric on C inherited from the norm $\|\cdot\|$,

$$\|x - y\| \leq d(x, y) \leq k \|x - y\| \quad \text{for all } x, y \in C.$$

Moreover, T is d-nonexpansive; that is,

$$d(Tx, Ty) \leq d(x, y) \quad \text{for any } x, y \in C.$$

On the other hand, suppose a metric ρ on C is equivalent to $\|\cdot\|$; that is, there exist constants $a, b > 0$ such that, for all $x, y \in C$,

$$a \|x - y\| \leq \rho(x, y) \leq b \|x - y\|. \tag{7.6}$$

Then, every ρ-nonexpansive mapping $T : C \to C$ satisfies

$$\|T^i x - T^i y\| \leq \frac{b}{a} \|x - y\|$$

for all $x, y \in C$ and $i = 1, 2, \dots$. The above observation leads us to an equivalent formulation of the constant $\gamma_0(X)$:

$\gamma_0(X) = \sup \{\, k \,:\, \text{any bounded, closed and convex set } C \subset X \text{ furnished}$

with a metric ρ satisfying (7.6) with $\dfrac{b}{a} < k$ has

the f.p.p. for ρ-nonexpansive mappings$\}$.

We have already seen in example 3.6 that, in every infinite dimensional Banach space $(X, \|\cdot\|)$, there is uniformly lipschitzian mapping $T : B_X \to B_X$, with a positive minimal displacement; that is,

$$d(T) = \inf \{\|x - Tx\| : x \in B_X\} > 0.$$

On the other hand, in the construction of our mapping T, we used a lipschitzian retraction $R : B_X \to S_X$ of the whole unit ball B_X onto its boundary S_X. Since every such retraction must have sufficiently large Lipschitz constant, this method does not lead us to nice upper bound for $\gamma_0(X)$.

An elegant evaluation from above for $\gamma_0(H)$ in the case of Hilbert space H is due to Baillon [4].

Example 7.11 (Baillon, 1978-79) *Consider the space ℓ_2 with the standard norm,*

$$\|x\| = \|(x_1, x_2, \dots)\| = \left(\sum_{i=1}^{\infty} x_i^2 \right)^{\frac{1}{2}}.$$

Let $T : \ell_2 \to \ell_2$ be the "right shift" operator; that is,

$$T(x_1, x_2, \dots) = (0, x_1, x_2, \dots).$$

Consider the positive part of the unit ball

$$B^+ = \{x \in \ell_2 : \|x\| \le 1 \text{ and } x_i \ge 0 \text{ for } i = 1, 2 \dots \}.$$

Define the mapping $F : B^+ \to B^+$ as

$$Fx = \begin{cases} \left(\cos \frac{\pi \|x\|}{2} \right) e_1 + \frac{1}{\|x\|} \left(\sin \frac{\pi \|x\|}{2} \right) Tx & \text{if } x \ne 0 \\ e_1 & \text{if } x = 0, \end{cases}$$

where $e_1 = (1, 0, \dots, 0, \dots)$. The verification that F is fixed point free as well as uniformly $\frac{\pi}{2}$-lipschitzian is left for the reader as an exercise.

Consequently, for a Hilbert space H, we have

$$\sqrt{2} \le \gamma_0(H) \le \frac{\pi}{2}.$$

The exact value of $\gamma_0(H)$ is still unknown!

Another approach, using the uniform normal structure coefficient, was presented by Casini and Maluta [19] in 1985.

Theorem 7.13 (Casini and Maluta) *Let C be a bounded, closed and convex subset of a Banach space X with $N(X) > 1$. If $T : C \to C$ is uniformly k-lipschitzian with*

$$k < \sqrt{N(X)},$$

then T has a fixed point.

One can observe that, for any Banach space X, we have $\kappa(X) \leq N(X)$. In the case of Hilbert space, the Lifshitz Theorem provides better lower bound for $\gamma_0(H)$ because $\kappa(H) = N(H) = \sqrt{2}$. On the other hand, there exist Banach spaces for which $\kappa(X) = 1$, whereas $N(X) > 1$. Hence, Casini-Maluta's Theorem and Lifshitz Theorem are not comparable.

In [8], Domínguez Benavides improved evaluation for $\gamma_0(X)$ obtained by Lifshitz. He introduced a new constant that involves both the normal structure coefficient and the Lifshitz constant.

Theorem 7.14 (Domínguez Benavides) *Let C be a bounded, closed and convex subset of a Banach space X. Let $T : C \to C$ be uniformly k-lipschitzian mapping. If*

$$k < \frac{1 + \sqrt{1 + 4N(X)(\kappa(X) - 1)}}{2},$$

then T has a fixed point.

Since $\kappa(X) \leq N(X)$, we immediately conclude that

$$\kappa(X) \leq \frac{1 + \sqrt{1 + 4N(X)(\kappa(X) - 1)}}{2} \leq N(X).$$

It is easy to verify that in the above inequality

$$\kappa(X) = \frac{1 + \sqrt{1 + 4N(X)(\kappa(X) - 1)}}{2}$$

if and only if $\kappa(X) = 1$ or $\kappa(X) = N(X)$. In particular, in the case of Hilbert space H, the Domínguez Benavides Theorem and the Lifshitz Theorem give the same lower bound for $\gamma_0(H)$, $\gamma_0(H) \geq \sqrt{2}$. However, in [8], we can find a class of Banach spaces with $1 < \kappa(X) < N(X)$. Obviously, for such spaces, the theorem of Domínguez Benavides is a strict improvement of Lifshitz's result.

Another approach to stability of f.p.p. uses only those equivalent metrics that are generated by equivalent norms. Recall that two norms $\|\cdot\|_1$ and $\|\cdot\|_2$ on a Banach space X are equivalent if there exist constants $a, b > 0$ such that inequality

$$a \|x\|_2 \leq \|x\|_1 \leq b \|x\|_2$$

holds for all $x \in X$. Then, every $\|\cdot\|_2$-nonexpansive mapping $T : C \to C$ is uniformly k-lipschitzian with respect to the norm $\|\cdot\|_1$ with $k \leq \frac{b}{a}$. Under this setting, we define the *stability constant* $\gamma_{\mathcal{N}}(X)$ of $(X, \|\cdot\|_1)$ as

$$\gamma_{\mathcal{N}}(X) = \sup \left\{ \gamma : \text{any Banach space } (X, \|\cdot\|_2) \text{ with } \frac{b}{a} < \gamma \text{ has the f.p.p.} \right.$$
$$\left. \text{for } \|\cdot\|_2\text{-nonexpansive mappings} \right\}.$$

It is clear that

$$\gamma_\mathcal{N}(X) \geq \gamma_0(X).$$

In [55], Lin proved that, for a Hilbert space H,

$$\gamma_\mathcal{N}(H) \geq \sqrt{\frac{5 + \sqrt{13}}{2}} = 2.07 \cdots > \frac{\pi}{2} \geq \gamma_0(H).$$

Later, this result was slightly improved by Mazcuñán-Navarro [61], with a conclusion

$$\gamma_\mathcal{N}(H) \geq \sqrt{\frac{5 + \sqrt{17}}{2}} = 2.135 \ldots .$$

In particular, the above result implies that the space ℓ_2 renormed as in example 7.9; that is,

$$|||x||| = \max \left\{ \frac{1}{\sqrt{2}} \|x\|, \max_{i \in \mathbb{N}} |x_i| \right\},$$

has the f.p.p. for $|||\cdot|||$-nonexpansive mappings because in this case $\frac{b}{a} = \sqrt{2}$.

7.5   The case of ℓ_1

The first result concerning lack of stability of f.p.p. for the space ℓ_1 was obtained by Goebel and Kuczumow [37] in 1978.

Let $\{a_i\}$ be a bounded sequence of nonnegative real numbers. If $\{e_i\}$ is a standard basis in ℓ_1, then

$$S^+ = \overline{\text{conv}}\, \{e_i : i \in \mathbb{N}\} = \left\{ x = \sum_{i=1}^{\infty} x_i e_i : x_i \geq 0 \text{ and } \sum_{i=1}^{\infty} x_i = 1 \right\}.$$

We shall modify the set S^+ by moving its vertexes e_i as follows. Let $f^i = (1 + a_i)\, e_i$ for $i \in \mathbb{N}$. Put $a = \inf a_i$ and $N_0 = \{i : a_i = a\}$. Let

$$C = \overline{\text{conv}}\, \{f^i : i \in \mathbb{N}\} = \left\{ x = \sum_{i=1}^{\infty} \lambda_i f^i : \lambda_i \geq 0 \text{ and } \sum_{i=1}^{\infty} \lambda_i = 1 \right\}.$$

It is easy to note that C is bounded, closed and convex. However, C is not weakly* compact in ℓ_1 considered as a dual of c_0 because the origin is the weak* limit of the sequence of vertexes f^i. The weak* closure $\overline{C}^{w*}$ of conv $(C \cup \{0\})$ has the form

$$\overline{C}^{w*} = \left\{ x = \sum_{i=1}^{\infty} \mu_i f^i : \mu_i \geq 0 \text{ and } \sum_{i=1}^{\infty} \mu_i \leq 1 \right\}.$$

Observe that a representation of x as a combination of f^i is unique. Hence, for each $x \in \overline{C}^{w*}$, we can define a number δ_x as

$$\delta_x = 1 - \sum_{i=1}^{\infty} \mu_i.$$

Then, for each $x \in \overline{C}^{w*}$,

$$\text{dist}\,(x, C) = (1 + a)\,\delta_x.$$

Indeed, since $x + \delta_x f^i \in C$, we have

$$
\begin{aligned}
\text{dist}\,(x, C) &\leq \inf \left\{ \left\| x - \left(x + \delta_x f^i \right) \right\| : i = 1, 2, \dots \right\} \\
&= \inf \left\{ \delta_x \left\| f^i \right\| : i = 1, 2, \dots \right\} \\
&= \inf \left\{ (1 + a_i)\,\delta_x : i = 1, 2, \dots \right\} \\
&= (1 + a)\,\delta_x.
\end{aligned}
$$

On the other hand, for any $y = \sum_{i=1}^{\infty} \lambda_i f^i \in C$,

$$
\begin{aligned}
\|x - y\| &= \sum_{i=1}^{\infty} |\mu_i - \lambda_i|\,(1 + a_i) \\
&\geq \left| \sum_{i=1}^{\infty} \lambda_i - \sum_{i=1}^{\infty} \mu_i \right| (1 + a) \\
&= \left(1 - \sum_{i=1}^{\infty} \mu_i \right) (1 + a) \\
&= (1 + a)\delta_x.
\end{aligned}
$$

Moreover, the above calculations imply that $\|x - y\| > (1 + a)\delta_x$, provided $y \in C \setminus \overline{\text{conv}} \left\{ x + \delta_x f^i : i \in N_0 \right\}$. Consequently,

$$P_C(x) := \{ z \in C : \|z - x\| = \text{dist}(x, C) \} = \overline{\text{conv}} \left\{ x + \delta_x f^i : i \in N_0 \right\}.$$

Case 1. N_0 is nonempty but finite. Let $\{x^n\}$ be an approximate fixed point sequence for a nonexpansive mapping $T : C \to C$. Without loss of generality, we can assume that $\{x^n\}$ is weak* convergent to $x \in \overline{C}^{w*} \subset \ell_1 = c_0^*$. Then,

$$A\left(\{x^n\}, C \right) = \text{conv} \left\{ x + \delta_x f^i : i \in N_0 \right\}$$

is compact, convex, and T-invariant; hence, T has a fixed point in $A\left(\{x^n\}, C \right) \subset C$.

Consequently, if N_0 is nonempty but finite, then C has f.p.p. for nonexpansive mappings.

Case 2. N_0 **is empty.** Then, there exists a subsequence $\{a_{i_k}\}_{k=1}^{\infty}$ of $\{a_i\}$ such that

$$\forall_{k\in\mathbb{N}}\forall_{i\in\mathbb{N}}\left(i < i_k \Rightarrow a_i > a_{i_k}\right).$$

It is clear that $\{a_{i_k}\}_{k=1}^{\infty}$ is strictly decreasing and $\lim\limits_{k\to\infty} a_{i_k} = a$. Put

$$N_1 = \{i : i \le i_1\}$$

and

$$N_k = \{i : i_{k-1} < i \le i_k\} \quad \text{for } k = 2, 3, \ldots.$$

Let

$$T : x = \sum_{i=1}^{\infty} \lambda_i f^i \to \sum_{k=1}^{\infty} \beta_k f^{i_{k+1}},$$

where

$$\beta_k = \sum_{i\in N_k} \lambda_i.$$

Since

$$\sum_{k=1}^{\infty} \beta_k = \sum_{k=1}^{\infty}\sum_{i\in N_k} \lambda_i = \sum_{i=1}^{\infty} \lambda_i = 1,$$

we conclude that $T : C \to C$. The mapping T is not only nonexpansive but even contractive. Indeed, for any $x = \sum_{i=1}^{\infty} \lambda_i f^i$ and $y = \sum_{i=1}^{\infty} \mu_i f^i$, $x \ne y$, we have

$$\|Tx - Ty\| = \left\|\sum_{k=1}^{\infty}\left\{\left[\sum_{i\in N_k}(\lambda_i - \mu_i)\right]f^{i_{k+1}}\right\}\right\|$$

$$\le \sum_{k=1}^{\infty}\sum_{i\in N_k}|\lambda_i - \mu_i|\left(1 + a_{i_{k+1}}\right)$$

$$< \sum_{k=1}^{\infty}\sum_{i\in N_k}|\lambda_i - \mu_i|\left(1 + a_i\right)$$

$$= \sum_{i=1}^{\infty}|\lambda_i - \mu_i|\left(1 + a_i\right)$$

$$= \|x - y\|.$$

Suppose that, for some $x = \sum_{i=1}^{\infty} \lambda_i f^i$, $Tx = x$. Then $\sum_{i\in N_k} \lambda_i = \lambda_{i_{k+1}}$ for each $k \in \mathbb{N}$. This implies that $\lambda_{i_k} \le \lambda_{i_{k+1}}$, and because $\{\lambda_{i_k}\}_{k=1}^{\infty}$ is subsequence of $\{\lambda_i\}_{i=1}^{\infty} \subset \ell_1$, we conclude that $\lambda_{i_k} = 0$ for each $k \in \mathbb{N}$. However, this contradicts the fact that

$$1 = \sum_{k=1}^{\infty}\sum_{i\in N_k} \lambda_i = \sum_{k=1}^{\infty} \lambda_{i_{k+1}}.$$

Thus, if N_0 is empty, then C fails the f.p.p. for nonexpansive (and even contractive) mappings!

Case 3. N_0 is infinite. Let $N_0 = \{i_1, i_2, \dots\}$. Let

$$N_1 = \{i : i \leq i_1\}$$

and

$$N_k = \{i : i_{k-1} < i \leq i_k\} \quad \text{for } k = 2, 3, \dots.$$

We leave to the reader the verification of the fact that the mapping $T : C \to C$ defined as in the previous case is nonexpansive and fixed point free.

Hence, if N_0 is infinite, then C fails the f.p.p. for nonexpansive mappings!

We can summarize the above cases as follows:

Theorem 7.15 (Goebel and Kuczumow, 1978) *Let C and N_0 be as above. Then, the set C has the f.p.p. if and only if N_0 is nonempty but finite.*

Subsequently, the stability of the fixed point property for nonexpansive self-mappings defined on bounded, closed and convex subsets of ℓ_1 has been widely studied by Dowling, Lennard, and Turett. The first step in their approach was the following:

Theorem 7.16 (Strong Version of James's Distortion Theorem, [22]) *A Banach space X contains an isomorphic copy of ℓ_1 if and only if, for each sequence $\{\epsilon_n\}_{n=1}^{\infty}$ in $(0,1)$ such that $\lim_{n \to \infty} \epsilon_n = 0$, there exists a sequence $\{x_n\}_{n=1}^{\infty}$ in X such that*

$$(1 - \epsilon_k) \sum_{n=k}^{\infty} |t_n| \leq \left\| \sum_{n=k}^{\infty} t_n x_n \right\| \leq \sum_{n=k}^{\infty} |t_n| \tag{7.7}$$

holds for all $\{t_n\}_{n=1}^{\infty} \in \ell_1$ and for all $k = 1, 2, \dots.$

Using the above, Dowling, Lennard, and Turett [24] proved the following

Theorem 7.17 *Let X be a Banach space containing an isomorphic copy of ℓ_1. Then, for each $\epsilon > 0$, there exists a bounded, closed and convex subset C of X and a uniformly lipschitzian mapping $T : C \to C$ that is fixed point free and $k_u(T) < 1 + \epsilon$.*

Proof. Fix $\epsilon > 0$. Let $\{\epsilon_n\}_{n=1}^{\infty}$ in $(0,1)$ be a null sequence such that

$$\frac{1}{1 - \epsilon_1} < 1 + \epsilon.$$

By the Strong Version of James's Distortion Theorem, there exists a sequence $\{x_n\}_{n=1}^{\infty}$ in X satisfying (7.7). Consider the set $C \subset \ell_1$ defined as

$$C = \left\{ \sum_{n=1}^{\infty} t_n x_n : t_n \geq 0 \text{ and } \sum_{n=1}^{\infty} t_n = 1 \right\}.$$

The set C is bounded, closed and convex subset of ℓ_1. Define the mapping $T : C \to C$ by

$$T \left(\sum_{n=1}^{\infty} t_n x_n \right) = \sum_{n=1}^{\infty} t_n x_{n+1}.$$

Then, for each $y = \sum_{n=1}^{\infty} t_n x_n$, $z = \sum_{n=1}^{\infty} s_n x_n$ in C and for any $k \in \mathbb{N}$,

$$
\begin{aligned}
\left\| T^k y - T^k z \right\| &= \left\| \sum_{n=1}^{\infty} (t_n - s_n) x_{n+k} \right\| \\
&\leq \sum_{n=1}^{\infty} |t_n - s_n| \\
&\leq \frac{1}{1 - \epsilon_1} \left\| \sum_{n=1}^{\infty} (t_n - s_n) x_n \right\| \\
&\leq \frac{1}{1 - \epsilon_1} \|y - z\|.
\end{aligned}
$$

Consequently,

$$k_u(T) \leq \frac{1}{1 - \epsilon_1} < 1 + \epsilon.$$

Obviously, T is fixed point free.

$\square$

It means that ℓ_1 cannot be renormed to have the fixed point property for uniformly lipschitzian mappings having uniform Lipschitz constant slightly greater than 1. Hence,

$$\gamma_{\mathcal{N}}(\ell_1) = \gamma_0(\ell_1) = 1.$$

Immediately, one can ask about renorming of ℓ_1 having the f.p.p. for non-expansive mappings. To proceed, we shall need the concept of **asymptotically isometric copy of** ℓ_1, which was initiated and developed by Dowling, Lennard, and Turett (see [49]).

A Banach space X is said to contain *an asymptotically isometric copy of* ℓ_1 if, for every null sequence $\{\epsilon_n\}_{n=1}^{\infty}$ in $(0, 1)$, there exists a sequence $\{x_n\}_{n=1}^{\infty}$ in X such that

$$\sum_{n=1}^{\infty} (1 - \epsilon_n) |t_n| \leq \left\| \sum_{n=1}^{\infty} t_n x_n \right\| \leq \sum_{n=1}^{\infty} |t_n|,$$

for all $\{t_n\}_{n=1}^{\infty} \in \ell_1$.

Theorem 7.18 (see [23] or [49]) *If a Banach space X contains an asymptotically isometric copy of ℓ_1, then X fails the f.p.p. for nonexpansive mappings.*

Proof. Let $\{\lambda_i\}_{i=1}^{\infty}$ be a strictly decreasing sequence in $(1, \infty)$ converging to 1. Then, there exists a sequence $\{\epsilon_i\}_{i=1}^{\infty}$ in $(0, 1)$ such that $\lim_{i \to \infty} \epsilon_i = 0$ and

$$\lambda_{i+1} < (1 - \epsilon_i)\lambda_i$$

for $i = 1, 2, \ldots$. By assumption, there exists a sequence $\{x_i\}_{i=1}^{\infty}$ in X such that

$$\sum_{i=1}^{\infty}(1 - \epsilon_i)\,|t_i| \leq \left\| \sum_{i=1}^{\infty} t_i x_i \right\| \leq \sum_{i=1}^{\infty} |t_i|$$

for every $\{t_i\}_{i=1}^{\infty} \in \ell_1$. Put $y_i = \lambda_i x_i$ for $i = 1, 2, \ldots$. Define the set $C \subset X$ as

$$C = \left\{ \sum_{i=1}^{\infty} t_i y_i : \{t_i\}_{i=1}^{\infty} \in \ell_1, t_i \geq 0 \text{ and } \sum_{i=1}^{\infty} t_i = 1 \right\}.$$

We leave to the reader the verification that C is a bounded, closed and convex subset of X.

Now, consider the mapping $T : C \to C$ defined by

$$T\left(\sum_{i=1}^{\infty} t_i y_i \right) = \sum_{i=1}^{\infty} t_i y_{i+1}.$$

It is easy to check that T is fixed point free. We shall prove that T is nonexpansive. For any $u = \sum_{i=1}^{\infty} t_i y_i$ and $v = \sum_{i=1}^{\infty} s_i y_i$ in C, we have

$$
\begin{aligned}
\|Tu - Tv\| &= \left\| \sum_{i=1}^{\infty} (t_i - s_i)\, y_{i+1} \right\| \\
&\leq \sum_{i=1}^{\infty} (t_i - s_i)\, \|y_{i+1}\| \\
&\leq \sum_{i=1}^{\infty} (t_i - s_i)\, \lambda_{i+1} \\
&\leq \sum_{i=1}^{\infty} (t_i - s_i)\, \lambda_i\, (1 - \epsilon_i) \\
&\leq \left\| \sum_{i=1}^{\infty} (t_i - s_i)\lambda_i x_i \right\| \\
&= \|u - v\|.
\end{aligned}
$$

$\square$

Theorem 7.19 (see [49]) *Let* Y *be a closed infinite-dimensional subspace of* ℓ_1 *furnished with the standard norm; that is,* $\|x\| = \sum_{i=1}^{\infty} |x_i|$ *for* $x = (x_1, x_2, \ldots) \in \ell_1$. *Then,* Y *contains an asymptotically isometric copy of* ℓ_1.

Theorem 7.20 (see [23] or [49]) *If* X *is a nonreflexive subspace of* $L_1[0,1]$, *then* X *contains an asymptotically isometric copy of* ℓ_1.

Hence, each nonreflexive subspace of $L_1[0,1]$ fails the f.p.p. for nonexpansive mappings. On the other hand, Maurey [60] proved that each reflexive subspace of $L_1[0,1]$ has the fixed point property. Consequently, we obtain the following

Theorem 7.21 *Let* X *be a subspace of* $L_1[0,1]$. *The following conditions are equivalent:*

(i) X *is reflexive.*

(ii) X *has the fixed point property for nonexpansive mappings.*

There exist renormings of ℓ_1 that fail to contain asymptotically isometric copies of ℓ_1.

Example 7.12 *[49] Consider the space* ℓ_1 *with a new norm defined by*

$$\|x\|_{DLT} = \|(x_1, x_2, \ldots)\|_{DLT} = \max\left\{ \gamma_i \sum_{k=i}^{\infty} |x_k| : i = 1, 2, \ldots \right\},$$

where $\{\gamma_i\}$ *is an arbitrary sequence in* $(0,1)$ *strictly increasing to* 1. *It is easy to note that, for any* $x \in \ell_1$

$$\gamma_1 \|x\|_{\ell_1} \leq \|x\|_{DLT} \leq \|x\|_{\ell_1}.$$

Dowling, Lennard, and Turett proved that $(\ell_1, \|\cdot\|_{DLT})$ *does not contain an asymptotically isometric copy of* ℓ_1. *Nevertheless, it was not clear whether there is* $\|\cdot\|_{DLT}$*-nonexpansive mapping that is fixed point free. To see it, consider, for instance, the well-known set*

$$S^+ = \left\{ \{x_i\} : x_i \geq 0 \text{ and } \sum_{i=1}^{\infty} x_i = 1 \right\}.$$

Obviously S^+ *is convex, and, since* $\|\cdot\|_{DLT}$ *is equivalent to the classical norm, it is also bounded and closed. Let* T *be the classical "right shift" self-mapping defined on* S^+,

$$T(x_1, x_2, \ldots) = (0, x_1, x_2, \ldots).$$

The mapping T *is fixed point free. However,* T *is not nonexpansive with respect to the new norm* $\|\cdot\|_{DLT}$ *with* $\gamma_i = \frac{8^i}{1+8^i}$ *for* $i = 1, 2, \ldots$. *We shall find the consecutive Lipschitz constants for iterates* T^n *of* T. *Since the sequence* $\left\{ \frac{\gamma_{i+1}}{\gamma_i} \right\}$

is strictly decreasing, for every $x = (x_1, x_2, \ldots)$, $y = (y_1, y_2, \ldots)$ *in* S^+, *and for all* $n \in \mathbb{N}$, *we have*

$$\|T^n x - T^n y\|_{DLT} = \max_{i \in \mathbb{N}} \gamma_{n+i} \sum_{k=i}^{\infty} |x_k - y_k|$$

$$= \max_{i \in \mathbb{N}} \frac{\gamma_{n+i}}{\gamma_{n+i-1}} \cdot \frac{\gamma_{n+i-1}}{\gamma_{n+i-2}} \cdot \ldots \cdot \frac{\gamma_{i+1}}{\gamma_i} \cdot \gamma_i \cdot \sum_{k=i}^{\infty} |x_k - y_k|$$

$$\leq \max_{i \in \mathbb{N}} \frac{\gamma_{n+1}}{\gamma_n} \cdot \frac{\gamma_n}{\gamma_{n-1}} \cdot \ldots \cdot \frac{\gamma_2}{\gamma_1} \cdot \gamma_i \cdot \sum_{k=i}^{\infty} |x_k - y_k|$$

$$= \frac{\gamma_{n+1}}{\gamma_1} \cdot \max_{i \in \mathbb{N}} \gamma_i \sum_{k=i}^{\infty} |x_k - y_k|$$

$$= \frac{\gamma_{n+1}}{\gamma_1} \|x - y\|_{DLT}.$$

Further,

$$\|T^n e_1 - T^n e_2\|_{DLT} = 2\gamma_{n+1}$$

and

$$\|e_1 - e_2\|_{DLT} = 2\gamma_1,$$

hence,

$$\|T^n e_1 - T^n e_2\|_{DLT} = \frac{\gamma_{n+1}}{\gamma_1} \|e_1 - e_2\|_{DLT}.$$

Consequently, $k_{\|\cdot\|_{DLT}}(T^n) = \frac{\gamma_{n+1}}{\gamma_1} \longrightarrow \frac{1}{\gamma_1} = \frac{9}{8}$ *as* $n \longrightarrow \infty$.

In 2008, Lin [56] proved that ℓ_1 furnished with the above-mentioned norm has the fixed point property for nonexpansive mappings!

Theorem 7.22 (Lin, 2008) *The space ℓ_1 furnished with an equivalent norm $\|\cdot\|_{DLT}$ defined as*

$$\|x\|_{DLT} = \|(x_1, x_2, \ldots)\|_{DLT} = \max \left\{ \gamma_i \sum_{k=i}^{\infty} |x_k| : i = 1, 2, \ldots \right\},$$

with $\gamma_i = \frac{8^i}{1+8^i}$ has the fixed point property for $\|\cdot\|_{DLT}$-nonexpansive mappings.

It was also the first example of nonreflexive Banach space with the fixed point property!

Reassuming bounded, closed and convex subsets of $(\ell_1, \|\cdot\|_{DLT})$ with $\gamma_i = \frac{8^i}{1+8^i}$ have the fixed point property for nonexpansive mappings but for any $\epsilon > 0$, there exists a bounded, closed and convex subset C of $(\ell_1, \|\cdot\|_{DLT})$ and $(1 + \epsilon)$-uniformly lipschitzian mapping $T : C \to C$, which is fixed point free.

Chapter 8

Mean nonexpansive mappings

Recall that, for any convex subset C of a Banach space X with $\text{diam}(C) > 0$ and for any multi-index $\alpha = (\alpha_1, \ldots, \alpha_n)$, there is an α-nonexpansive mapping $T : C \to C$ that is not nonexpansive. We also know that a mapping T is α-nonexpansive if and only if it is nonexpansive with respect to the metric d defined by

$$d(x,y) = \sum_{j=1}^{n} \left(\sum_{i=j}^{n} \alpha_i \right) \left\| T^{j-1}x - T^{j-1}y \right\|.$$

Moreover, every time we prove the fixed point theorem for nonexpansive mappings, we get a fixed point theorem for some class of α-nonexpansive mappings.

Consequently, the fixed point property for nonexpansive mappings is always stable with respect to special changes of metrics. In this chapter, we shall develop the idea concerning this new kind of stability.

8.1 Some new results of stability type

As usual, C denotes a bounded closed and convex subset of a Banach space $(X, \|\cdot\|)$ and $\alpha = (\alpha_1, \ldots, \alpha_n)$ a multi-index. Fix $p \geq 1$.

Under the above settings, we say that a mapping $T : C \to C$ is (α, p)-nonexpansive if for all $x, y \in C$

$$\sum_{i=1}^{n} \alpha_i \left\| T^i x - T^i y \right\|^p \leq \|x - y\|^p. \tag{8.1}$$

Equivalently, $T : C \to C$ is (α, p)-nonexpansive if and only if it is nonexpansive with respect to the metric d defined for all $x, y \in C$ as

$$d(x,y) = \left[\sum_{j=1}^{n} \left(\sum_{i=j}^{n} \alpha_i \right) \left\| T^{j-1}x - T^{j-1}y \right\|^p \right]^{\frac{1}{p}}. \tag{8.2}$$

It is easy to see that d is equivalent to the metric ρ on C generated by the norm, $\rho(x, y) = \|x - y\|$ for all $x, y \in C$. Further, for each (α, p)-nonexpansive mapping $T : C \to C$ we define the mapping $T_\alpha : C \to C$ by putting

$$T_\alpha x = \sum_{i=1}^{n} \alpha_i T^i x \quad \text{for any } x \in C.$$

Since every (α, p)-nonexpansive mapping T is $(\alpha, 1)$-nonexpansive, we conclude that T_α is nonexpansive. Thus, $d(T_\alpha) = 0$.

Theorem 8.1 *Let C be a bounded, closed and convex subset of a Banach space X. If $T : C \to C$ is (α, p)-nonexpansive with $\alpha = (\alpha_1, \alpha_2)$ and $p \geq 1$ such that $\alpha_2^p + \alpha_2 \leq 1$, then $d(T) = 0$.*

Proof. If $p = 1$, then the conclusion follows from theorem 3.2. Suppose $p > 1$. Fix $\epsilon > 0$. Since T_α is nonexpansive, there exists a point $x \subset C$, $x = x(\epsilon)$, such that
$$\|x - T_\alpha x\| = \|x - \alpha_1 T x - \alpha_2 T^2 x\| < \epsilon.$$
By definition of T, we have

$$
\begin{aligned}
\alpha_1 \left\|T x - T^2 x\right\|^p + \alpha_2 \left\|T^2 x - T^3 x\right\|^p &\leq \|x - T x\|^p \\
&\leq (\|x - T_\alpha x\| + \|T_\alpha x - T x\|)^p \\
&\leq (\|T_\alpha x - T x\| + \epsilon)^p \\
&= \left(\alpha_2 \left\|T^2 x - T x\right\| + \epsilon\right)^p.
\end{aligned}
$$

Let $q > 1$ be such that $\frac{1}{p} + \frac{1}{q} = 1$. Using Hölder inequality, we obtain

$$
\begin{aligned}
\left(\alpha_2 \left\|T^2 x - T x\right\| + \epsilon\right)^p &= \left(\alpha_2 \left\|T^2 x - T x\right\| + \epsilon^{\frac{1}{q}} \cdot \epsilon^{\frac{1}{p}}\right)^p \\
&\leq \left[(1 + \epsilon)^{\frac{1}{q}} \cdot \left(\alpha_2^p \left\|T^2 x - T x\right\|^p + \epsilon\right)^{\frac{1}{p}}\right]^p \\
&= (1 + \epsilon)^{p-1} \left(\alpha_2^p \left\|T^2 x - T x\right\|^p + \epsilon\right) \\
&= \alpha_2^p \left\|T^2 x - T x\right\|^p + \left[(1 + \epsilon)^{p-1} - 1\right] \alpha_2^p \left\|T^2 x - T x\right\|^p \\
&\quad + \epsilon(1 + \epsilon)^{p-1} \\
&\leq \alpha_2^p \left\|T^2 x - T x\right\|^p + \left[(1 + \epsilon)^{p-1} - 1\right] (\operatorname{diam}(C))^p \\
&\quad + \epsilon(1 + \epsilon)^{p-1}.
\end{aligned}
$$

Thus,

$$
\begin{aligned}
(\alpha_1 - \alpha_2^p) \left\|T x - T^2 x\right\|^p + \alpha_2 \left\|T^2 x - T^3 x\right\|^p &\leq \left[(1 + \epsilon)^{p-1} - 1\right] (\operatorname{diam}(C))^p \\
&\quad + \epsilon(1 + \epsilon)^{p-1}.
\end{aligned}
$$

The right-hand side of the above inequality can be made arbitrary close to zero for small ϵ. By assumption,

$$\alpha_1 - \alpha_2^p = 1 - \alpha_2 - \alpha_2^p \geq 0.$$

Thus, $d(T) = 0$.

$\square$

If C has the f.p.p. for nonexpansive mappings, then we can repeat arguments from the previous theorem with $\epsilon = 0$ to get the following:

Theorem 8.2 *Let C be a bounded, closed and convex subset of a Banach space X. If C has the f.p.p. for nonexpansive mappings, then C has the f.p.p. for (α, p)-nonexpansive mappings with $\alpha = (\alpha_1, \alpha_2)$ and $p \geq 1$ such that $\alpha_2^p + \alpha_2 \leq 1$.*
   In other words, if for some $\alpha = (\alpha_1, \alpha_2)$ and $p \geq 1$ such that $\alpha_2^p + \alpha_2 \leq 1$, a mapping $T : C \to C$ is nonexpansive with respect to the metric d defined for all $x, y \in C$ by

$$d(x, y) = (\|x - y\|^p + \alpha_2 \|Tx - Ty\|^p)^{\frac{1}{p}},$$

then T has a fixed point.
   Moreover, if $\alpha_2^p + \alpha_2 < 1$, then $\mathrm{Fix}(T) = \mathrm{Fix}(T_\alpha)$.

Proof. Since T_α is nonexpansive, there is $z \in C$ such that

$$T_\alpha z = \alpha_1 Tz + \alpha_2 T^2 z = z. \tag{8.3}$$

By definition of T, we have

$$\alpha_1 \left\|Tz - T^2 z\right\|^p + \alpha_2 \left\|T^2 z - T^3 z\right\|^p \leq \|z - Tz\|^p = \|T_\alpha z - Tz\|^p$$
$$= \alpha_2^p \left\|T^2 z - Tz\right\|^p.$$

Thus,

$$(\alpha_1 - \alpha_2^p) \left\|Tz - T^2 z\right\|^p + \alpha_2 \left\|T^2 z - T^3 z\right\|^p \leq 0.$$

If $\alpha_1 - \alpha_2^p = 1 - \alpha_2 - \alpha_2^p > 0$, then $Tz = T^2 z$ and from (8.3) we get $Tz = z$. Thus, $\mathrm{Fix}(T_\alpha) \subset \mathrm{Fix}(T)$. Inclusion in opposite direction is obvious.
   If $\alpha_1 - \alpha_2^p = 0$, then $T^2 z$ is a fixed point of T.

$\square$

For the convenience of the reader, we illustrate the set of all pairs (p, α_2) for which the above theorems hold using figure 8.1.
   Let us observe that, for any $\alpha_2 \in (0, 1)$, there exists $p \geq 1$ such that C has the f.p.p. for all mappings in $L((\alpha_1, \alpha_2), p)$. Nevertheless, the above theorem

does not guarantee that there exists $p \geq 1$ such that for every $\alpha_2 \in (0,1)$, C has the f.p.p. for all mappings in $L((\alpha_1, \alpha_2), p)$.

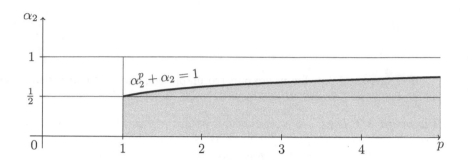

FIGURE 8.1

Similar situation is observed in the general case of $((\alpha_1, \ldots, \alpha_n), p)$-nonexpansive mappings as seen in the following theorems:

Theorem 8.3 *If $T : C \to C$ is (α, p)-nonexpansive with $n \geq 2$, $\alpha = (\alpha_1, \ldots, \alpha_n)$ and $p \geq 1$ such that*

$$(1 - \alpha_1)(1 - \alpha_1^{\frac{n-1}{p}}) \leq \alpha_1^{\frac{n-1}{p}}(1 - \alpha_1^{\frac{1}{p}}),$$

then $d(T) = 0$.

Proof. If $p = 1$, then the conclusion follows from theorem 3.4. Suppose $p > 1$. For every $\epsilon > 0$, there exists $x \in C$ such that $\|x - T_\alpha x\| \leq \epsilon$. Then,

$$\left(\sum_{i=1}^{n} \alpha_i \left\| T^i x - T^{i+1} x \right\|^p \right)^{1/p} \leq \|x - Tx\| \leq \|x - T_\alpha x\| + \|T_\alpha x - Tx\|$$

$$\leq \left\| \sum_{i=1}^{n} \alpha_i T^i x - \sum_{i=1}^{n} \alpha_i Tx \right\| + \epsilon$$

$$= \left\| \sum_{i=2}^{n} \alpha_i \left(T^i x - Tx \right) \right\| + \epsilon$$

$$\leq \sum_{i=2}^{n} \alpha_i \left\| T^i x - Tx \right\| + \epsilon.$$

Since $k(T^j) \leq \left(\frac{1}{\alpha_1^{1/p}} \right)^j$, we get

$$\alpha_i \left\| T^i x - Tx \right\| = \alpha_i \left\| \sum_{j=2}^{i} \left(T^{j-1} x - T^j x \right) \right\|$$

$$\leq \alpha_i \sum_{j=2}^{i} \left\| T^{j-1}x - T^j x \right\|$$

$$\leq \left[\alpha_i \sum_{j=2}^{i} \left(\frac{1}{\alpha_1^{1/p}} \right)^{j-2} \right] \left\| Tx - T^2 x \right\|.$$

Thus,

$$\sum_{i=2}^{n} \alpha_i \left\| T^i x - Tx \right\| \leq \left[\sum_{i=2}^{n} \sum_{j=2}^{i} \frac{\alpha_i}{\alpha_1^{(j-2)/p}} \right] \left\| Tx - T^2 x \right\|$$

$$= \left\{ \sum_{j=2}^{n} \left[\left(\sum_{i=j}^{n} \alpha_i \right) \frac{1}{\alpha_1^{(j-2)/p}} \right] \right\} \left\| Tx - T^2 x \right\|.$$

Since

$$\sum_{i=j}^{n} \alpha_i \leq 1 - \alpha_1$$

for $j = 2, \ldots, n$, we obtain

$$\sum_{i=2}^{n} \alpha_i \left\| T^i x - Tx \right\| \leq (1 - \alpha_1) \left(\sum_{j=2}^{n} \frac{1}{\alpha_1^{(j-2)/p}} \right) \left\| Tx - T^2 x \right\|$$

$$= (1 - \alpha_1) \cdot \frac{1}{\alpha_1^{(n-2)/p}} \cdot \left(\sum_{j=2}^{n} \alpha_1^{(j-2)/p} \right) \left\| Tx - T^2 x \right\|$$

$$= (1 - \alpha_1) \cdot \frac{1}{\alpha_1^{(n-2)/p}} \cdot \frac{1 - \alpha_1^{(n-1)/p}}{1 - \alpha_1^{1/p}} \left\| Tx - T^2 x \right\|.$$

If we define A as

$$A = (1 - \alpha_1) \cdot \frac{1}{\alpha_1^{(n-2)/p}} \cdot \frac{1 - \alpha_1^{(n-1)/p}}{1 - \alpha_1^{1/p}},$$

then we can write

$$\left(\sum_{i=1}^{n} \alpha_i \left\| T^i x - T^{i+1} x \right\|^p \right)^{1/p} \leq A \left\| Tx - T^2 x \right\| + \epsilon.$$

Thus,

$$\sum_{i=1}^{n} \alpha_i \left\| T^i x - T^{i+1} x \right\|^p \leq \left(A \left\| Tx - T^2 x \right\| + \epsilon \right)^p. \qquad (8.4)$$

We can select $q > 1$ such that $\frac{1}{p} + \frac{1}{q} = 1$. Using Hölder inequality, we obtain

$$
\begin{aligned}
\left(A \left\| Tx - T^2x \right\| + \epsilon \right)^p &= \left(A \left\| Tx - T^2x \right\| + \epsilon^{\frac{1}{p}} \cdot \epsilon^{\frac{1}{q}} \right)^p \\
&\leq \left[\left(A^p \left\| Tx - T^2x \right\|^p + \epsilon \right)^{\frac{1}{p}} \cdot (1 + \epsilon)^{\frac{1}{q}} \right]^p \\
&= \left(A^p \left\| Tx - T^2x \right\|^p + \epsilon \right) (1 + \epsilon)^{p-1} \\
&= A^p \left\| Tx - T^2x \right\|^p + \left[(1 + \epsilon)^{p-1} - 1 \right] A^p \left\| Tx - T^2x \right\|^p \\
&\quad + \epsilon (1 + \epsilon)^{p-1} \\
&\leq A^p \left\| Tx - T^2x \right\|^p + \left[(1 + \epsilon)^{p-1} - 1 \right] A^p \left(\mathrm{diam}(C) \right)^p \\
&\quad + \epsilon (1 + \epsilon)^{p-1} .
\end{aligned}
$$

Thus,

$$
(\alpha_1 - A^p) \left\| Tx - T^2x \right\|^p
$$
$$
+ \sum_{i=2}^{n} \alpha_i \left\| T^i x - T^{i+1}x \right\|^p \leq \left[(1 + \epsilon)^{p-1} - 1 \right] A^p \left(\mathrm{diam}(C) \right)^p
$$
$$
+ \epsilon (1 + \epsilon)^{p-1} .
$$

By assumption $\alpha_1 - A^p \geq 0$, this implies that

$$
\alpha_n \left\| T^n x - T^{n+1}x \right\|^p \leq \left[(1 + \epsilon)^{p-1} - 1 \right] A^p \left(\mathrm{diam}(C) \right)^p + \epsilon (1 + \epsilon)^{p-1} .
$$

Since $\alpha_n > 0$ and the right-hand side of the above inequality converges to 0 if $\epsilon \to 0$, we finally obtain $d(T) = 0$.

$\square$

If C has the f.p.p. for nonexpansive mappings, then we can repeat the proof with $\epsilon = 0$ to obtain the following:

Theorem 8.4 *If C has the f.p.p. for nonexpansive mappings, then C has the f.p.p. for (α, p)-nonexpansive mappings with $n \geq 2$, $\alpha = (\alpha_1, \ldots, \alpha_n)$ and $p \geq 1$ such that*
$$
(1 - \alpha_1)(1 - \alpha_1^{\frac{n-1}{p}}) \leq \alpha_1^{\frac{n-1}{p}} (1 - \alpha_1^{\frac{1}{p}}).
$$

Equivalently, if for some α and p satisfying the above condition, a mapping $T : C \to C$ is nonexpansive with respect to the metric d given by formula (8.2), then T has a fixed point.

Moreover, if
$$
(1 - \alpha_1)(1 - \alpha_1^{\frac{n-1}{p}}) < \alpha_1^{\frac{n-1}{p}} (1 - \alpha_1^{\frac{1}{p}}),
$$

then $Fix(T) = Fix(T_\alpha)$.

Proof. If $p = 1$, then the first conclusion follows from theorem 3.5. Suppose $p > 1$. Since $T_\alpha : C \to C$ is nonexpansive, there exists $x \in C$ such that

$$T_\alpha x = \alpha_1 Tx + \alpha_2 T^2 x + \cdots + \alpha_n T^n x = x. \tag{8.5}$$

Now inequality (8.4) takes the form

$$\sum_{i=1}^n \alpha_i \left\| T^i x - T^{i+1} x \right\|^p \le A^p \left\| Tx - T^2 x \right\|^p.$$

Observe that the above inequality also holds for $p = 1$. Thus,

$$(\alpha_1 - A^p) \left\| Tx - T^2 x \right\|^p + \sum_{i=2}^n \alpha_i \left\| T^i x - T^{i+1} x \right\|^p \le 0.$$

Since $\alpha_1 - A^p \ge 0$ and $\alpha_n > 0$, this implies that $T^n x$ is fixed under T. If $\alpha_1 - A^p > 0$, then

$$Tx = T^2 x = \cdots = T^n x.$$

From (8.5), we get that $x = Tx$. Thus, $\text{Fix}(T_\alpha) \subset \text{Fix}(T)$. Inclusion in opposite direction is obvious.

□

The reader can easily verify that for $p = 1$ the above theorem coincides with the theorem 3.5.

In view of theorems 8.4, 7.5, and 7.6, we obtain the following:

Theorem 8.5 *Let $(X, \|\cdot\|)$ be a Banach space that is uniformly nonsquare or uniformly noncreasy. Then, any bounded, closed and convex subset C has the f.p.p. for (α, p)-nonexpansive mappings with $n \ge 2$, $\alpha = (\alpha_1, \ldots, \alpha_n)$, and $p \ge 1$ such that*

$$(1 - \alpha_1)(1 - \alpha_1^{\frac{n-1}{p}}) \le \alpha_1^{\frac{n-1}{p}} (1 - \alpha_1^{\frac{1}{p}}).$$

Moreover, if

$$(1 - \alpha_1)(1 - \alpha_1^{\frac{n-1}{p}}) < \alpha_1^{\frac{n-1}{p}} (1 - \alpha_1^{\frac{1}{p}}),$$

then $\text{Fix}(T) = \text{Fix}(T_\alpha) \ne \emptyset$.

In particular, for $n = 2$, the above theorem states:

Theorem 8.6 *Let $(X, \|\cdot\|)$ be a Banach space that is uniformly nonsquare or uniformly noncreasy. Then, any bounded, closed and convex subset C has the f.p.p. for (α, p)-nonexpansive mappings with $\alpha = (\alpha_1, \alpha_2)$ and $p \ge 1$ such that $\alpha_2^p + \alpha_2 \le 1$. Moreover, if $\alpha_2^p + \alpha_2 < 1$, then $\text{Fix}(T) = \text{Fix}(T_\alpha) \ne \emptyset$.*

For the space ℓ_1, we have $\epsilon_0(\ell_1) = 2$. Nevertheless, weak* compact and convex sets in $\ell_1 = c_0^*$ have the f.p.p. for nonexpansive mappings.

Theorem 8.7 *Weak* compact and convex sets in ℓ_1 (considered as a dual of c_0) have the f.p.p. for (α, p)-nonexpansive mappings with $n \geq 2$, $\alpha = (\alpha_1, \ldots, \alpha_n)$, and $p \geq 1$ such that*

$$(1 - \alpha_1)(1 - \alpha_1^{\frac{n-1}{p}}) \leq \alpha_1^{\frac{n-1}{p}} (1 - \alpha_1^{\frac{1}{p}}).$$

Moreover, if

$$(1 - \alpha_1)(1 - \alpha_1^{\frac{n-1}{p}}) < \alpha_1^{\frac{n-1}{p}} (1 - \alpha_1^{\frac{1}{p}}),$$

then $Fix(T) = Fix(T_\alpha) \neq \emptyset$.

For $n = 2$, the above theorem states:

Theorem 8.8 *Weak* compact and convex sets in $\ell_1 = c_0^*$ have the f.p.p. for (α, p)-nonexpansive mappings with $\alpha = (\alpha_1, \alpha_2)$ and $p \geq 1$ such that $\alpha_2^p + \alpha_2 \leq 1$. Furthermore, if $\alpha_2^p + \alpha_2 < 1$, then $Fix(T) = Fix(T_\alpha) \neq \emptyset$.*

In view of Lin's Theorem, we obtain the following result:

Theorem 8.9 *Consider the space ℓ_1 furnished with the norm $\|\cdot\|_{DLT}$ defined as*

$$\|x\|_{DLT} = \|(x_1, x_2, \ldots)\|_{DLT} = \max \left\{ \gamma_n \sum_{k=n}^{\infty} |x_k| : n = 1, 2, \ldots \right\},$$

where $\gamma_n = \frac{8^n}{1+8^n}$ for $n = 1, 2, \ldots$. Let C be a bounded, closed and convex subset of ℓ_1 and $T : C \to C$ be a mapping. If for some $\alpha = (\alpha_1, \ldots, \alpha_n)$ and $p \geq 1$ satisfying

$$(1 - \alpha_1)(1 - \alpha_1^{\frac{n-1}{p}}) \leq \alpha_1^{\frac{n-1}{p}} (1 - \alpha_1^{\frac{1}{p}}),$$

the mapping T is nonexpansive with respect to the metric d given by

$$d(x, y) = \left[\sum_{j=1}^{n} \left(\sum_{i=j}^{n} \alpha_i \right) \|T^{j-1}x - T^{j-1}y\|_{DLT}^p \right]^{\frac{1}{p}},$$

then T has a fixed point.

Another approach uses the constant $k_\infty(T)$. As was observed in corollary 5.2 and theorem 5.4, for every (α, p)-nonexpansive mapping T with $\alpha = (\alpha_1, \alpha_2)$ and $p \geq 1$,

$$k_\infty(T) \leq (1 + \alpha_2)^{1/p},$$

and, in a general case of $((\alpha_1, \ldots, \alpha_n), p)$-nonexpansive mappings,

$$k_\infty(T) \leq \left(\sum_{j=1}^{n} \left(\sum_{i=j}^{n} \alpha_i \right) \right)^{1/p} = (\alpha_1 + 2\alpha_2 + 3\alpha_3 + \cdots + n\alpha_n)^{1/p}$$

$$= (1 + \alpha_2 + 2\alpha_3 + \cdots + (n-1)\alpha_n)^{1/p}.$$

Using the above inequality and the Lifshitz Theorem, we conclude

Theorem 8.10 (Pérez García and Piasecki, [65]) *Let* (M, ρ) *be a bounded and complete metric space with* $\kappa(M) > 1$. *If* $T : M \to M$ *is* (α, p)-*nonexpansive with* $p \geq 1$ *and* $\alpha = (\alpha_1, \ldots, \alpha_n)$ *such that*

$$(1 + \alpha_2 + 2\alpha_3 + \cdots + (n-1)\alpha_n)^{1/p} < \kappa(M),$$

then T *has a fixed point in* M.

Now we are ready to prove the following:

Theorem 8.11 (Pérez García and Piasecki, [65]) *Let* X *be a Banach space with* $\epsilon_0(X) < 1$. *If* $T : C \to C$ *is* (α, p)-*nonexpansive with* $n \geq 2$, $p \geq 1$ *and* $\alpha = (\alpha_1, \ldots, \alpha_n)$ *such that*

$$(1 + \alpha_2 + 2\alpha_3 + \cdots + (n-1)\alpha_n)^{1/p} < \kappa(X),$$

then T *has a fixed point in* C.

Proof. It is a consequence of theorems 7.12, 5.4, and 7.11.

$\square$

Hence, for $n = 2$, we have

Theorem 8.12 *Let* X *be a Banach space with* $\epsilon_0(X) < 1$. *If* $T : C \to C$ *is* (α, p)-*nonexpansive with* $\alpha = (\alpha_1, \alpha_2)$ *and* $p \geq 1$ *such that*

$$(1 + \alpha_2)^{1/p} < \kappa(X),$$

then T *has a fixed point in* C.

In particular, since $\kappa(H) = \sqrt{2}$ for a Hilbert space H, theorem 8.11 states:

Theorem 8.13 *Let* H *be a Hilbert space. If* $T : C \to C$ *is* (α, p)-*nonexpansive, with* $n \geq 2$, $p \geq 1$, *and* $\alpha = (\alpha_1, \ldots, \alpha_n)$ *such that*

$$1 + \alpha_2 + 2\alpha_3 + \cdots + (n-1)\alpha_n < 2^{p/2},$$

then T *has a fixed point in* C.

For $n = 2$, we get

Theorem 8.14 *Let* H *be a Hilbert space. If* $T : C \to C$ *is* (α, p)-*nonexpansive, with* $p \geq 1$ *and* $\alpha = (\alpha_1, \alpha_2)$ *such that*

$$\alpha_2 < 2^{p/2} - 1,$$

then T *has a fixed point in* C.

For mappings in the class $L(\alpha, 2, 1)$, the theorem 8.13 takes the form

Theorem 8.15 *Let H be a Hilbert space. If $T : C \to C$ is $(\alpha, 2)$-nonexpansive, with $n \geq 2$ and $\alpha = (\alpha_1, \ldots, \alpha_n)$ such that*

$$\alpha_2 + 2\alpha_3 + 3\alpha_4 + \cdots + (n-1)\alpha_n < 1,$$

then T has a fixed point in C.

Let us note that, for $n = 2$, the last theorem implies:

Theorem 8.16 *Let C be a bounded, closed and convex subset of a Hilbert space H. Then, for any multi-index $\alpha = (\alpha_1, \alpha_2)$ of length $n = 2$, C has the f.p.p. for all $(\alpha, 2)$-nonexpansive mappings.*
In other words, if for some $\alpha = (\alpha_1, \alpha_2)$, a mapping $T : C \to C$ is nonexpansive with respect to the metric d given by

$$d(x,y) = \left(\|x - y\|^2 + \alpha_2 \|Tx - Ty\|^2 \right)^{\frac{1}{2}},$$

then T has a fixed point.

In general, if $\epsilon_0(X) < 1$, then for any $n \geq 2$, there exists sufficiently large p such that for any multi-index α of length n, C has the f.p.p. for all mappings in the class $L(\alpha, p, 1)$. This fact is a consequence of the following theorem:

Theorem 8.17 (Pérez García and Piasecki, [65]) *Let C be a bounded, closed and convex subset of a Banach space X with $\epsilon_0(X) < 1$. Then, for any multi-index $\alpha = (\alpha_1, \ldots, \alpha_n)$ of length $n \geq 2$ and $p = \frac{\ln(n)}{\ln(\kappa(X))}$, C has the f.p.p. for the class of (α, p)-nonexpansive mappings.*
In other words, if for some $\alpha = (\alpha_1, \ldots, \alpha_n)$, a mapping $T : C \to C$ is nonexpansive with respect to the metric d given by

$$d(x,y) = \left[\sum_{j=1}^{n} \left(\sum_{i=j}^{n} \alpha_i \right) \|T^{j-1}x - T^{j-1}y\|^p \right]^{\frac{1}{p}},$$

with $p = \frac{\ln(n)}{\ln(\kappa(X))}$, then T has a fixed point.

Proof. It is enough to observe that, for any $\alpha = (\alpha_1, \ldots, \alpha_n)$, we have

$$\sum_{i=1}^{n} \alpha_i = 1 \quad \text{and} \quad \sum_{i=j}^{n} \alpha_i < 1 \quad \text{for} \quad j = 2, \ldots, n.$$

Thus,

$$\left(\sum_{j=1}^{n} \left(\sum_{i=j}^{n} \alpha_i \right) \right)^{1/p} < n^{1/p} = \kappa(X).$$

$\square$

Consequently, in case of Hilbert space, we have

Theorem 8.18 *If C is a bounded, closed and convex subset of a Hilbert space H, then for any multi-index* $\alpha = (\alpha_1, \ldots, \alpha_n)$ *of length* $n \geq 2$ *and* $p = \frac{2 \ln(n)}{\ln(2)}$, *C has the f.p.p. for the class of* (α, p)*-nonexpansive mappings.*

In other words, if $T : C \to C$ *is a mapping such that for some* $\alpha = (\alpha_1, \ldots, \alpha_n)$ *it is nonexpansive with respect to the metric d defined as*

$$d(x, y) = \left[\sum_{j=1}^{n} \left(\sum_{i=j}^{n} \alpha_i \right) \left\| T^{j-1}x - T^{j-1}y \right\|^{\frac{2\ln(n)}{\ln(2)}} \right]^{\frac{\ln(2)}{2\ln(n)}},$$

then T has a fixed point.

Now, we shall compare the approach that uses the constants $k_\infty(T)$ and $\kappa(X)$ with the previous one, which involved the mapping T_α. For technical reasons, we consider the case of (α, p)-nonexpansive mappings with a multi-index α of length $n = 2$. Hence, we shall compare theorem 8.12 with theorem 8.6.

If $\epsilon_0(X) \geq 1$, then the situation is clear. Only theorem 8.6 provides some new fixed point results for the class of mean nonexpansive mappings. However, if $\epsilon_0(X) < 1$, then the situation is much more complicated. For clarity of arguments, we consider the case of Hilbert space H. Then, theorem 8.12 takes the form of theorem 8.14.

We have already noticed that for $p = 2$ theorem 8.14 ensures the f.p.p. for the whole class of $(\alpha, 2)$-nonexpansive mappings, whereas, using theorem 8.6, we obtain the f.p.p. only for $(\alpha, 2)$-nonexpansive mappings with $\alpha_2 \leq \frac{\sqrt{5}-1}{2}$. On the other hand, for $p = 1$, theorem 8.14 ensures the f.p.p. for all α-nonexpansive mappings with $\alpha_2 < \sqrt{2} - 1$, whereas using theorem 8.6, we get the f.p.p. for all α-nonexpansive mappings with $\alpha_2 \leq \frac{1}{2}$, which is a better estimate.

To display more precisely the above remark, we use figure 8.2. By $E = (p', \alpha_2') = (1.240..., 0.537...)$, we denote the point at which graphs of functions $\alpha_2^p + \alpha_2 = 1$ (see theorem 8.6) and $\alpha_2 = 2^{p/2} - 1$ (see theorem 8.14) intersect. It is easy to observe that, for $p \leq p'$, theorem 8.6 gives better estimate for α_2 than theorem 8.14, whereas for $p > p'$ the situation is contrary.

We summarize the above observation in the following:

Theorem 8.19 *Let C be a bounded, closed and convex subset of a Hilbert space H. If* $T : C \to C$ *is* (α, p)*-nonexpansive, with* $\alpha = (\alpha_1, \alpha_2)$ *and* $p \geq 1$ *such that*

$$p \in [1, p'] \ \text{and} \ \alpha_2^p + \alpha_2 \leq 1$$

or

$$p > p' \ \text{and} \ \alpha_2 < \min\left\{ 2^{\frac{p}{2}} - 1, 1 \right\},$$

then T has a fixed point in C.

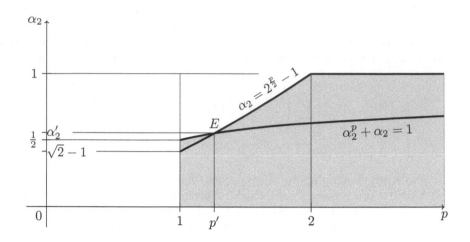

FIGURE 8.2

8.2   Sequential approximation of fixed points

Let C be a bounded, closed and convex subset of uniformly convex Banach space X. Let $\alpha = (\alpha_1, \ldots, \alpha_n)$ and p be such that (see theorem 8.4):

$$(1 - \alpha_1)(1 - \alpha_1^{\frac{n-1}{p}}) < \alpha_1^{\frac{n-1}{p}}(1 - \alpha_1^{\frac{1}{p}}).$$

Then, in view of theorem 8.4 and the proof of theorem 7.1, we can give a procedure to determine the fixed point for (α, p)-nonexpansive mapping $T : C \to C$ (α and p as above).

1. Construct an approximate fixed point sequence $\{x_m\}$ for mapping T_α as in the proof of lemma 3.1 with $\epsilon_m = \frac{1}{m}$ for $m = 1, 2, \ldots$.

2. Determine the asymptotic center of the sequence $\{x_m\}$, which consists of exactly one point z.

Then, z is a fixed point of T.

Actually, under the above assumption on α and p, the sets of approximate fixed point sequences for mappings T and T_α coincide, as seen in the following theorems:

Theorem 8.20 *Let C be a nonempty and convex subset of a Banach space X. If $T : C \to C$ is (α, p)-nonexpansive with $\alpha = (\alpha_1, \alpha_2)$ and $p \geq 1$ such that $\alpha_2^p + \alpha_2 < 1$, then for each sequence $\{x_n\} \subset C$,*

$$\lim_{n\to\infty} \|x_n - Tx_n\| = 0 \iff \lim_{n\to\infty} \|x_n - T_\alpha x_n\| = 0.$$

Proof. Suppose $\lim_{n\to\infty} \|x_n - Tx_n\| = 0$. Then,

$$
\begin{aligned}
\|T_\alpha x_n - x_n\| &\leq \alpha_1 \|Tx_n - x_n\| + \alpha_2 \|T^2 x_n - x_n\| \\
&\leq \alpha_1 \|Tx_n - x_n\| + \alpha_2 \|T^2 x_n - Tx_n\| + \alpha_2 \|Tx_n - x_n\| \\
&= \|Tx_n - x_n\| + \alpha_2 \|T^2 x_n - Tx_n\| \\
&\leq \|Tx_n - x_n\| + \frac{\alpha_2}{\alpha_1^{1/p}} \|Tx_n - x_n\| \\
&= \left(1 + \frac{\alpha_2}{\alpha_1^{1/p}}\right) \|Tx_n - x_n\|.
\end{aligned}
$$

Thus, $\lim_{n\to\infty} \|T_\alpha x_n - x_n\| = 0$.

Conversely, let $\lim_{n\to\infty} \|T_\alpha x_n - x_n\| = 0$. Then,

$$
\begin{aligned}
\|x_n - Tx_n\| &\leq \|x_n - T_\alpha x_n\| + \|T_\alpha x_n - Tx_n\| \\
&= \|x_n - T_\alpha x_n\| + \alpha_2 \|T^2 x_n - Tx_n\| \\
&\leq \|x_n - T_\alpha x_n\| + \frac{\alpha_2}{\alpha_1^{1/p}} \|Tx_n - x_n\|.
\end{aligned}
$$

Shifting the last term to the left-hand side and multiplying both sides by $\alpha_1^{1/p}$ we obtain

$$\left(\alpha_1^{1/p} - \alpha_2\right) \|Tx_n - x_n\| \leq \alpha_1^{1/p} \|x_n - T_\alpha x_n\|.$$

It is clear that our assumption $\alpha_2^p + \alpha_2 < 1$ is equivalent to $\alpha_1^{1/p} - \alpha_2 > 0$. Hence, we conclude that $\lim_{n\to\infty} \|Tx_n - x_n\| = 0$.

$\square$

In the general case of $((\alpha_1, \ldots, \alpha_n), p)$-nonexpansive mappings, we have the following:

Theorem 8.21 *Let C be a nonempty and convex subset of a Banach space X. If $T : C \to C$ is (α, p)-nonexpansive with $n \geq 2$, $\alpha = (\alpha_1, \ldots, \alpha_n)$, and $p \geq 1$ such that*

$$(1 - \alpha_1)(1 - \alpha_1^{\frac{n-1}{p}}) < \alpha_1^{\frac{n-1}{p}} (1 - \alpha_1^{\frac{1}{p}}),$$

then for any sequence $\{x_m\}$ in C,

$$\lim_{m\to\infty} \|x_m - Tx_m\| = 0 \iff \lim_{m\to\infty} \|x_m - T_\alpha x_m\| = 0.$$

Proof. Suppose $\lim_{m \to \infty} \|x_m - Tx_m\| = 0$. Then,

$$\|x_m - T_\alpha x_m\| = \left\|\sum_{i=1}^{n} \alpha_i \left(x_m - T^i x_m\right)\right\| \le \sum_{i=1}^{n} \alpha_i \left\|x_m - T^i x_m\right\|.$$

Since $k(T^j) \le 1/\alpha_1^{j/p}$, for $i = 1, \ldots, n$, we obtain

$$\|x_m - T^i x_m\| = \left\|\sum_{j=0}^{i-1}(T^j x_m - T^{j+1} x_m)\right\| \le \sum_{j=0}^{i-1} \left\|T^j x_m - T^{j+1} x_m\right\|$$

$$\le \left(\sum_{j=0}^{i-1} 1/\alpha_1^{j/p}\right) \|x_m - Tx_m\|.$$

Thus,

$$\|x_m - T_\alpha x_m\| \le \left[\sum_{i=1}^{n} \left(\sum_{j=0}^{i-1} \frac{\alpha_i}{\alpha_1^{j/p}}\right)\right] \|x_m - Tx_m\|.$$

This implies that $\lim_{m \to \infty} \|x_m - T_\alpha x_m\| = 0$.
Inversely, suppose $\lim_{m \to \infty} \|x_m - T_\alpha x_m\| = 0$. Then,

$$\|x_m - Tx_m\| \le \|x_m - T_\alpha x_m\| + \|T_\alpha x_m - Tx_m\|$$

$$= \|x_m - T_\alpha x_m\| + \left\|\sum_{i=2}^{n} \alpha_i \left(T^i x_m - Tx_m\right)\right\|$$

$$\le \|x_m - T_\alpha x_m\| + \sum_{i=2}^{n} \alpha_i \left\|T^i x_m - Tx_m\right\|.$$

Since $k(T^j) \le 1/\alpha_1^{j/p}$, for $i = 2, \ldots, n$, we have

$$\left\|T^i x_m - Tx_m\right\| = \left\|\sum_{j=1}^{i-1}(T^j x_m - T^{j+1} x_m)\right\| \le \sum_{j=1}^{i-1} \left\|T^j x_m - T^{j+1} x_m\right\|$$

$$\le \left(\sum_{j=1}^{i-1} \frac{1}{\alpha_1^{j/p}}\right) \|x_m - Tx_m\|.$$

Thus,

$$\sum_{i=2}^{n} \alpha_i \left\|T^i x_m - Tx_m\right\| \le \left[\sum_{i=2}^{n} \left(\sum_{j=1}^{i-1} \frac{\alpha_i}{\alpha_1^{j/p}}\right)\right] \|x_m - Tx_m\|.$$

Adding terms with the same denominator, we get

$$\sum_{i=2}^{n}\left(\sum_{j=1}^{i-1}\frac{\alpha_i}{\alpha_1^{j/p}}\right)=\sum_{j=1}^{n-1}\left(\sum_{i=j+1}^{n}\frac{\alpha_i}{\alpha_1^{j/p}}\right).$$

Since

$$\sum_{i=j+1}^{n}\alpha_i\leq 1-\alpha_1,$$

we obtain

$$\sum_{j=1}^{n-1}\left(\sum_{i=j+1}^{n}\frac{\alpha_i}{\alpha_1^{j/p}}\right)\leq\sum_{j=1}^{n-1}\frac{1-\alpha_1}{\alpha_1^{j/p}}=\frac{(1-\alpha_1)}{\alpha_1^{(n-1)/p}}\left(\sum_{j=1}^{n-1}\alpha_1^{\frac{j-1}{p}}\right)$$

$$=(1-\alpha_1)\left(1-\alpha_1^{\frac{n-1}{p}}\right)\left[\alpha_1^{\frac{n-1}{p}}\left(1-\alpha_1^{\frac{1}{p}}\right)\right]^{-1}.$$

Thus,

$$\|x_m-Tx_m\|\leq\|x_m-T_\alpha x_m\|$$
$$+(1-\alpha_1)\left(1-\alpha_1^{\frac{n-1}{p}}\right)\left[\alpha_1^{\frac{n-1}{p}}\left(1-\alpha_1^{\frac{1}{p}}\right)\right]^{-1}\|x_m-Tx_m\|.$$

By assumption,

$$(1-\alpha_1)\left(1-\alpha_1^{\frac{n-1}{p}}\right)\left[\alpha_1^{\frac{n-1}{p}}\left(1-\alpha_1^{\frac{1}{p}}\right)\right]^{-1}<1.$$

Hence, $\lim_{m\to\infty}\|x_m-Tx_m\|=0$.

$\square$

Actually, evaluations based only on α_1 are not exact. In the next section, we shall illustrate this fact for a multi-index α of length $n=3$ and $p=1$.

8.3   The case of $n=3$

Let C be a bounded, closed and convex subset of a Banach space $(X,\|\cdot\|)$. Let $T:C\to C$ be α-nonexpansive with $\alpha=(\alpha_1,\alpha_2,\alpha_3)$; that is, the inequality

$$\alpha_1\|Tx-Ty\|+\alpha_2\left\|T^2x-T^2y\right\|+\alpha_3\left\|T^3x-T^3y\right\|\leq\|x-y\| \quad (8.6)$$

holds for all x, $y\in C$. Then, the mapping $T_\alpha:C\to C$ has the form

$$T_\alpha x=\alpha_1Tx+\alpha_2T^2x+\alpha_3T^3x.$$

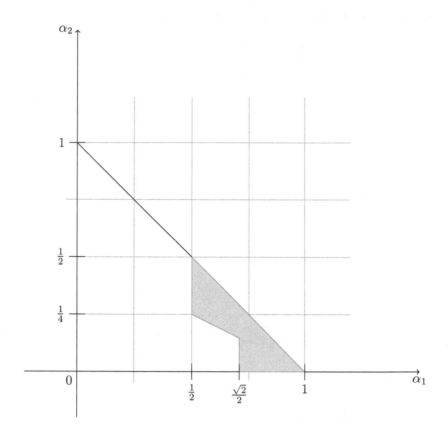

FIGURE 8.3

In [33], Goebel and Japón Pineda proved the following:

Theorem 8.22 *Let C be a bounded, closed and convex subset of a Banach space X having the f.p.p. for nonexpansive mappings. Suppose that a mapping $T : C \to C$ is α-nonexpansive with a multi-index $\alpha = (\alpha_1, \alpha_2, \alpha_3)$ such that*

$$\alpha_1 \in \left[\frac{1}{2}, \frac{\sqrt{2}}{2} \right) \quad and \quad \alpha_2 \geq \frac{1}{2} - \frac{1}{2}\alpha_1$$

or

$$\alpha_1 \geq \frac{\sqrt{2}}{2}.$$

Then, T has a fixed point in C (look at the figure 8.3).

Proof. It is enough to repeat the proof of the theorem 3.7, with $\epsilon = 0$.

$\square$

Looking up the proof of this theorem, we conclude that the evaluation for coefficients of $\alpha = (\alpha_1, \alpha_2, \alpha_3)$, which guarantees the f.p.p. can be improved as seen below.

Theorem 8.23 *Let C be a bounded, closed and convex subset of a Banach space X. If $T : C \to C$ is α-nonexpansive with $\alpha = (\alpha_1, \alpha_2, \alpha_3)$ such that $\alpha_1 \geq \frac{1}{2}$ and $\alpha_2 \geq \frac{1}{2} - \alpha_1^2$, then $d(T) = 0$.*

Proof. Fix $\epsilon > 0$. Since T_α is nonexpansive, there is a point $x \in C$ such that

$$\|x - T_\alpha x\| < \alpha_3 \epsilon.$$

Thus, using definition of T, we can write

$$\sum_{i=1}^{3} \alpha_i \left\| T^i x - T^{i+1} x \right\| \leq \|x - Tx\| \leq \|x - T_\alpha x\| + \|T_\alpha x - Tx\|$$

$$\leq \|T_\alpha x - Tx\| + \alpha_3 \epsilon$$

$$= \left\| \alpha_2 \left(T^2 x - Tx \right) + \alpha_3 \left(T^3 x - Tx \right) \right\| + \alpha_3 \epsilon$$

$$\leq \alpha_2 \left\| T^2 x - Tx \right\| + \alpha_3 \left\| T^3 x - Tx \right\| + \alpha_3 \epsilon$$

$$\leq \alpha_2 \left\| T^2 x - Tx \right\| + \alpha_3 \left\| T^3 x - T^2 x \right\|$$

$$+ \alpha_3 \left\| T^2 x - Tx \right\| + \alpha_3 \epsilon$$

$$= (\alpha_2 + \alpha_3) \left\| T^2 x - Tx \right\| + \alpha_3 \left\| T^3 x - T^2 x \right\| + \alpha_3 \epsilon.$$

Leaving only the last term on the left hand-side and shifting all others to the right we obtain

$$\alpha_3 \left\| T^3 x - T^4 x \right\| \leq (\alpha_2 + \alpha_3 - \alpha_1) \left\| T^2 x - Tx \right\|$$

$$+ (\alpha_3 - \alpha_2) \left\| T^3 x - T^2 x \right\| + \alpha_3 \epsilon$$

$$\leq (1 - 2\alpha_1) \left\| T^2 x - Tx \right\| + (\alpha_3 - \alpha_2) \left\| T^3 x - T^2 x \right\| + \alpha_3 \epsilon.$$

If $\alpha_1 \geq 1/2$ and $\alpha_2 \geq \alpha_3$, then

$$\left\| T^3 x - T^4 x \right\| \leq \epsilon.$$

Suppose $\alpha_2 < \alpha_3$. Since $k(T) \leq \frac{1}{\alpha_1}$, we get

$$\alpha_3 \left\| T^3 x - T^4 x \right\| \leq (1 - 2\alpha_1) \left\| T^2 x - Tx \right\| + \frac{\alpha_3 - \alpha_2}{\alpha_1} \left\| T^2 x - Tx \right\| + \alpha_3 \epsilon$$

$$= \frac{1 - 2\alpha_2 - 2\alpha_1^2}{\alpha_1} \left\| T^2 x - Tx \right\| + \alpha_3 \epsilon.$$

If $1 - 2\alpha_2 - 2\alpha_1^2 \leq 0$, then $\left\| T^3 x - T^4 x \right\| \leq \epsilon$. Thus, $d(T) = 0$.

$\square$

To compare the theorem 8.22 with the theorem 8.23 we present figure 8.4.

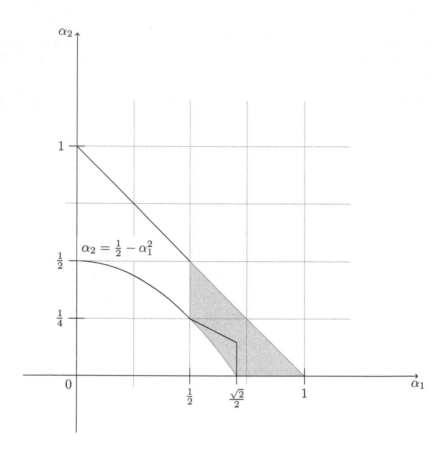

FIGURE 8.4

If the mapping T_α has a fixed point, $x = T_\alpha x$, then we can repeat the above proof with $\epsilon = 0$. Thus, we have the following:

Theorem 8.24 *Let C be a bounded, closed and convex subset of a Banach space X. If C has the f.p.p. for nonexpansive mappings, then C has the f.p.p. for α-nonexpansive mappings with $\alpha = (\alpha_1, \alpha_2, \alpha_3)$ such that $\alpha_1 \geq \frac{1}{2}$ and $\alpha_2 \geq \frac{1}{2} - \alpha_1^2$.*

In other words, if for some multi-index $\alpha = (\alpha_1, \alpha_2, \alpha_3)$ satisfying the above conditions, a mapping $T : C \to C$ is nonexpansive with respect to the metric d defined by

$$d(x, y) = \|x - y\| + (\alpha_2 + \alpha_3)\, \|Tx - Ty\| + \alpha_3 \,\|T^2x - T^2y\| ,$$

then T has a fixed point in C.

To present some further results related to asymptotic center methods, we recall some basic evaluations for Lipschitz constants of iterates of T. It is clear that, for all $x, y \in C$, we have

$$\|Tx - Ty\| \leq \frac{1}{\alpha_1} \|x - y\|. \tag{8.7}$$

Further, using (8.6), we can write

$$\alpha_2 \|T^2x - T^2y\| \leq \|x - y\| - \alpha_1 \|Tx - Ty\|$$
$$\leq \|x - y\| - \alpha_1^2 \|T^2x - T^2y\|.$$

Putting the last term on the left-hand side, we obtain

$$(\alpha_1^2 + \alpha_2) \|T^2x - T^2y\| \leq \|x - y\|.$$

Thus,

$$\|T^2x - T^2y\| \leq \frac{1}{\alpha_1^2 + \alpha_2} \|x - y\|. \tag{8.8}$$

Now we are ready to prove the following lemmas:

Lemma 8.1 *Let C be a nonempty and convex subset of a Banach space X. If $T : C \to C$ is α-nonexpansive with $\alpha = (\alpha_1, \alpha_2, \alpha_3)$ such that $\alpha_1 > \frac{1}{2}$ and $\alpha_2 > \frac{1}{2} - \alpha_1^2$, then $\mathrm{Fix}(T) = \mathrm{Fix}(T_\alpha)$.*

Proof. Obviously, $\mathrm{Fix}(T) \subset \mathrm{Fix}(T_\alpha)$. Clearly, we can assume that $\mathrm{Fix}(T_\alpha) \neq \emptyset$. Let $x \in \mathrm{Fix}(T_\alpha)$. Then,

$$\begin{aligned}
\|x - Tx\| &= \|T_\alpha x - Tx\| \\
&= \|\alpha_2 (T^2x - Tx) + \alpha_3 (T^3x - Tx)\| \\
&\leq \alpha_2 \|T^2x - Tx\| + \alpha_3 \|T^3x - T^2x\| + \alpha_3 \|T^2x - Tx\| \\
&\leq (\alpha_2 + \alpha_3) \|T^2x - Tx\| + \alpha_3 \|T^3x - T^2x\|.
\end{aligned}$$

If $\alpha_2 > 0$, then combining (8.7), (8.8), and α-nonexpansivness of T, we can write

$$\|x - Tx\| \leq \max\left\{(\alpha_2 + \alpha_3)s + \alpha_3 \frac{1 - \alpha_1 s}{\alpha_2} : s \in \left[\frac{\alpha_1}{\alpha_1^2 + \alpha_2}, \frac{1}{\alpha_1}\right]\right\} \|x - Tx\|$$

$$= \max\left\{\frac{\alpha_1^2 + \alpha_2 - \alpha_1}{\alpha_2}s + \frac{\alpha_3}{\alpha_2} : s \in \left[\frac{\alpha_1}{\alpha_1^2 + \alpha_2}, \frac{1}{\alpha_1}\right]\right\} \|x - Tx\|.$$

If $\alpha_1^2 + \alpha_2 - \alpha_1 \geq 0$, then the maximum is achieved for $s = 1/\alpha_1$. Thus,

$$\|x - Tx\| \leq \frac{1 - \alpha_1}{\alpha_1} \|x - Tx\|,$$

and for $\alpha_1 > \frac{1}{2}$, we get $x \in \mathrm{Fix}(T)$.

If $\alpha_1^2 + \alpha_2 - \alpha_1 < 0$, then the maximum is achieved for $s = \alpha_1/(\alpha_1^2 + \alpha_2)$. Thus,

$$\|x - Tx\| \leq \frac{1 - \alpha_1^2 - \alpha_2}{\alpha_1^2 + \alpha_2} \|x - Tx\| = \left(\frac{1}{\alpha_1^2 + \alpha_2} - 1\right) \|x - Tx\|,$$

and for $\alpha_2 > \frac{1}{2} - \alpha_1^2$, we have $x \in \mathrm{Fix}(T)$.

For $\alpha_2 = 0$, we can use (8.7) and (8.8) to obtain

$$\|x - Tx\| \leq \alpha_3 \left(\|T^2x - Tx\| + \|T^3x - T^2x\|\right) \leq \left(\frac{\alpha_3}{\alpha_1} + \frac{\alpha_3}{\alpha_1^2}\right) \|x - Tx\|$$

$$= \frac{1 - \alpha_1^2}{\alpha_1^2} \|x - Tx\|.$$

If $\alpha_1 > \frac{\sqrt{2}}{2}$, then $x \in \mathrm{Fix}(T)$.

All the above means that, for $\alpha_1 > \frac{1}{2}$ and $\alpha_2 > \frac{1}{2} - \alpha_1^2$, we have $\mathrm{Fix}(T) = \mathrm{Fix}(T_\alpha)$.

□

Lemma 8.2 *Let C be a nonempty and convex subset of a Banach space X. If $T : C \to C$ is α-nonexpansive with $\alpha = (\alpha_1, \alpha_2, \alpha_3)$ such that $\alpha_1 > \frac{1}{2}$ and $\alpha_2 > \frac{1}{2} - \alpha_1^2$, then for each sequence $\{x_n\} \subset C$,*

$$\lim_{n\to\infty} \|x_n - Tx_n\| = 0 \iff \lim_{n\to\infty} \|x_n - T_\alpha x_n\| = 0.$$

Proof. Suppose $\lim_{n\to\infty} \|x_n - Tx_n\| = 0$. Then, as in the proof of theorem 8.21, $\lim_{n\to\infty} \|T_\alpha x_n - x_n\| = 0$.

Conversely, let $\lim_{n\to\infty} \|T_\alpha x_n - x_n\| = 0$. Then,

$$\|x_n - Tx_n\| \leq \|x_n - T_\alpha x_n\| + \|T_\alpha x_n - Tx_n\|$$
$$\leq \|x_n - T_\alpha x_n\| + \alpha_2 \|T^2 x_n - Tx_n\| + \alpha_3 \|T^3 x_n - Tx_n\|$$
$$\leq \|x_n - T_\alpha x_n\| + (\alpha_2 + \alpha_3) \|T^2 x_n - Tx_n\|$$
$$+ \alpha_3 \|T^3 x_n - T^2 x_n\|.$$

Now, it is enough to repeat arguments from the previous lemma to get the conclusion.

□

Finally, we get the following theorem, which gives us sequential methods of approximation of fixed points:

Theorem 8.25 *Let C be a bounded closed and convex subset of a uniformly convex Banach space X. If $T : C \to C$ is α-nonexpansive with $\alpha = (\alpha_1, \alpha_2, \alpha_3)$ such that $\alpha_1 > \frac{1}{2}$ and $\alpha_2 > \frac{1}{2} - \alpha_1^2$, then the set of all a.f.p.s. for T coincides with the set of all a.f.p.s. for T_α and for each such sequence $\{x_n\}$, $A(C, \{x_n\})$ consists of exactly one point, which is fixed under T.*

Proof. It is a consequence of lemmas 8.1, 8.2, and the proof of theorem 7.1.

$\square$

All presented results can be generalized for $n \geq 4$, and difficulties are based only on some technicalities.

8.4   On the structure of the fixed points set

It is well known that, in the case of strictly convex space, the fixed points set of nonexpansive mapping is convex (it may be empty):

Theorem 8.26 *Let C be a nonempty, closed and convex subset of a strictly convex Banach space X. If $T : C \to C$ is nonexpansive, then $Fix(T)$ is closed and convex.*

Thus, in view of theorem 8.4, we have the following:

Conclusion 8.1 *Let C be a bounded, closed and convex subset of a strictly convex Banach space X. If $T : C \to C$ is (α, p)-nonexpansive with $\alpha - (\alpha_1, \dots, \alpha_n)$ such that*

$$(1 - \alpha_1)(1 - \alpha_1^{\frac{n-1}{p}}) < \alpha_1^{\frac{n-1}{p}}(1 - \alpha_1^{\frac{1}{p}}),$$

then $Fix(T)$ is closed and convex (it may be empty).

Actually, the result for nonexpansive mappings can be generalized to the whole class of α-nonexpansive ones.

Theorem 8.27 (Pérez García and Piasecki, [65]) *Let C be a nonempty, closed and convex subset of a strictly convex Banach space X. For any (α, p)-nonexpansive mapping $T : C \to C$ with $\alpha = (\alpha_1, \dots, \alpha_n)$ and $p \geq 1$, $Fix(T)$ is closed and convex (it may be empty).*

Proof. Since C is closed and T is continuous, $Fix(T)$ is closed in X. Obviously, it is enough to consider the case of α-nonexpansive mapping.

Suppose that $Fix(T)$ consists of more than one point. Let $x, y \in Fix(T)$, $x \neq y$. Then, $T([x, y]) \subset [x, y]$, where

$$[x, y] = \{(1 - t)x + ty : t \in [0, 1]\}.$$

Indeed, suppose there exists $z \in [x, y]$ such that $Tz \notin [x, y]$. Since X is strictly convex and T is α-nonexpansive, we obtain

$$\|x - y\| = \sum_{i=1}^{n} \alpha_i \|T^i x - T^i y\| < \sum_{i=1}^{n} \alpha_i \left(\|T^i x - T^i z\| + \|T^i z - T^i y\|\right)$$

$$\leq \sum_{i=1}^{n} \alpha_i \|T^i x - T^i z\| + \sum_{i=1}^{n} \alpha_i \|T^i z - T^i y\|$$

$$\leq \|x - z\| + \|z - y\| = \|x - y\|,$$

which is a contradiction. Since $T([x, y]) \subset [x, y]$, $Tx = x$, $Ty = y$, and T is continuous, we conclude that $T([x, y]) = [x, y]$.

Since $\mathrm{Fix}(T)$ is closed, it is enough to show that $\mathrm{Fix}(T)$ is dense in $[x, y]$. Suppose that $\mathrm{Fix}(T)$ is not dense in $[x, y]$. Without loss of generality, we can assume that there is not any fixed point in (x, y). Let $\varphi : [0, 1] \to [x, y]$ be a parametrization of $[x, y]$; that is,

$$\varphi(t) = (1 - t)x + ty.$$

It is clear that φ is bijection, φ and φ^{-1} are continuous. Let $f : [0, 1] \to [0, 1]$ be defined by

$$f = \varphi^{-1} T \varphi.$$

Since f is continuous and $\mathrm{Fix}(f) = \{0, 1\}$, this implies that either $f(t) > t$ for every $t \in (0, 1)$ or $f(t) < t$ for every $t \in (0, 1)$. Suppose that $f(t) > t$ for any $t \in (0, 1)$. Then, for every $z \in (x, y)$, we have $\|x - T^i z\| > \|x - z\|$, $i = 1, \dots, n$. Thus,

$$\sum_{i=1}^{n} \alpha_i \|T^i x - T^i z\| = \sum_{i=1}^{n} \alpha_i \|x - T^i z\| > \|x - z\|,$$

which contradicts α-nonexpansiveness of T. Thus, $\mathrm{Fix}(T)$ is dense in $[x, y]$. Since $\mathrm{Fix}(T)$ is closed, we conclude that $[x, y] \subset \mathrm{Fix}(T)$.

$\square$

If the space is not strictly convex, then the thesis of theorem 8.27 does not hold. To prove it, we use an example of nonexpansive mapping presented in [36], pp. 35, ex. 3.7:

Example 8.1 *In the space c_0 of sequences of real numbers with the limit 0 and standard sup norm let us consider a closed unit ball B and a mapping $T : B \to B$ defined by*

$$Tx = T(x_1, x_2, \dots) = (x_1, 1 - \|x\|, x_2, x_3, \dots).$$

It is easy to check that the mapping T is an isometry; that is, for each $x, y \in B$

$$\|Tx - Ty\| = \|x - y\|.$$

This implies that, for any α and $p \geq 1$, T is (α, p)-nonexpansive. It is easy to verify that the set of fixed points of T consists of two points $e_1 = (1, 0, \ldots, 0, \ldots)$ and $-e_1 = (-1, 0, \ldots, 0, \ldots)$, hence, is not convex.

Even if X is strictly convex, then the fixed points set for uniformly lipschitzian mapping may be not convex:

Example 8.2 *Let X be an infinite dimensional Banach space. Then there exists a retraction $R : B_X \to S_X$ of the closed unit ball B_X onto its boundary S_X satisfying the Lipschitz condition with a certain Lipschitz constant $k(R) = k \geq 3$. Since $R^n = R$ for $n \geq 1$, we conclude that R is uniformly lipschitzian with $k(R^n) = k$. In spite of this, $Fix(R) = S_X$; hence, is not convex. This in particular implies that R is not mean nonexpansive at least in case of strictly convex space.*

Actually, it is easy to verify that a retraction R is lipschitzian with $k(R) = k$ if and only if for any α, it is α-lipschitzian with $k(\alpha, R) = k$.

Chapter 9

Mean lipschitzian mappings with $k > 1$

The minimal displacement and *optimal retraction* problems have their roots in efforts to extend the celebrated Brouwer's Fixed Point Theorem to non-compact settings. Below, we briefly outline the evolution of this problem. We will consider stages that presumably have exerted the greatest impact on the development of this issue as well as some of the intriguing questions that were arising as the problem was confronted with more and more answers.

Then, we will widely study lipschitzian retractions onto a ball, lipschitzian self-mappings T defined on a bounded, closed, convex but noncompact subset C of infinite dimensional Banach space X with a positive minimal displacement, that is, $\inf \{ \|x - Tx\| : x \in C \} > 0$, as well as lipschitzian retractions of the whole unit ball onto its boundary with relatively small Lipschitz constants. This topic, which among specialists is called *the minimal displacement* and *optimal retraction problem*, has been widely studied in books [36], [31], and [49]. In spite of this, within the last 10 years, we have had a big progress. In this chapter, we shall give an updated presentation of this topic. Then, we shall extend obtained results to classes $L(\alpha, p, k)$ with $k > 1$.

9.1 Losing compactness in Brouwer's Fixed Point Theorem

The most famous and the most frequently quoted theorems in the topological fixed point theory were formulated by Brouwer in 1912 and Schauder in 1930. Before those fundamental findings are cited, let us start by presenting some classical definitions and facts:

A topological space X is said to have *the topological fixed point property* (generally abbreviated as t.f.p.p.), if each continuous mapping $f : X \to X$ has a fixed point. This property is topologically invariant. Indeed, suppose Y is a topological space, and $h : X \to Y$ is a homeomorphism such that $h(X) = Y$. Consider any continuous function $f : Y \to Y$. Then, the mapping $g = h^{-1} \circ f \circ h : X \to X$ is continuous. Thus, there exists a point $x \in X$ such that $(h^{-1} \circ f \circ h) x = x$, and, consequently, $f(h(x)) = h(x)$. This means that $h(x)$ is a fixed point of f. Thus, all sets homeomorphic to X share the topological fixed point property.

Let us recall that a subset $D \subset X$ is called *a retract* if there exists a continuous mapping (*a retraction*) $r : X \to D$ such that Fix$r = D$. Again, consider any continuous function $f : D \to D$, and extend it to the continuous mapping $g = f \circ r : X \to D \subset X$. It is easy to observe that Fix$(g) = $ Fix(f). Consequently, if X has the t.f.p.p., then this property is also inherited by all retracts of X.

In the sequel, we shall assume that $\mathbb{R}^n$ is furnished with the inner product defined for $x = (x_1, \ldots, x_n)$, $y = (y_1, \ldots, y_n)$ as $(x, y) = \sum_{i=1}^{n} x_i y_i$ and the norm $\|x\| = \left(\sum_{i=1}^{n} x_i^2\right)^{1/2}$. Moreover, we use B^n and S^{n-1} to denote the unit ball and the unit sphere in $\mathbb{R}^n$.

Within our terminology, Brouwer's Theorem might be formulated as follows [16]:

Theorem 9.1 (Brouwer's Fixed Point Theorem) *The unit ball $B^n \subset \mathbb{R}^n$ has the topological fixed point property.*

Obviously, every closed ball in $\mathbb{R}^n$ has the t.f.p.p., as it is homeomorphic to the unit ball B^n.

Suppose now that $C \subset \mathbb{R}^n$ is bounded, closed and convex. Then, in view of lemma 9.2, C is a nonexpansive retract of the whole space $\mathbb{R}^n$ via *the nearest point projection* $P_C : \mathbb{R}^n \to C$, which maps each point $x \in \mathbb{R}^n$ to the closest point $P_C x$ in C; that is, $\|P_C x - x\| = \inf \{\|x - y\| : y \in C\}$. Since C is bounded, it is contained in some closed ball in $\mathbb{R}^n$, and hence, it is a retract of that ball. This implies that all such sets have the topological fixed point property.

Finally, let us point out that every Banach space X of finite dimension is isomorphic to $\mathbb{R}^n$ for $n = \dim X$. In the light of the foregoing, Brouwer's theorem may be formulated in an equivalent yet structurally more general form:

Theorem 9.2 (Brouwer's Fixed Point Theorem) *Any bounded, closed and convex subset of a finite dimensional Banach space has the topological fixed point property.*

Presently, Brouwer's theorem may take many equivalent forms. Many information on various types of equivalents can be found in [34] as well as in [31] and [36]. We shall elaborate on two of them.

Theorem 9.3 (No retraction theorem) S^{n-1} *is not the retract of B^n.*

The equivalence of these two facts follows from the following observations. First, suppose that $R : B^n \to S^{n-1}$ would be a retraction of the unit ball B^n onto its boundary S^{n-1}. Then, the mapping $T = -R : B^n \to B^n$ is continuous and fixed point free. On the other hand, suppose that there exists a mapping

$T = B^n \to B^n$ without fixed points. Then, we can extend this mapping to the ball $2B^n$ by putting

$$Fx = \begin{cases} Tx & \text{if } \|x\| \le 1, \\ (2 - \|x\|)\, T\left(\frac{x}{\|x\|}\right) & \text{if } \|x\| \in [1, 2] \end{cases}$$

and observe that $F : 2B^n \to B^n$, F is continuous, fixed point free, and maps the doubled unit sphere $2S^{n-1}$ into the origin. By putting $\widetilde{F}x = \frac{1}{2} F(2x)$ we obtain a continuous, fixed point free mapping $\widetilde{F} : B^n \to B^n$ such that $\widetilde{F}(S^{n-1}) = \{0\}$, and, consequently, the retraction $R : B^n \to S^{n-1}$,

$$Rx = \frac{x - \widetilde{F}x}{\left\|x - \widetilde{F}x\right\|}.$$

Our next equivalent variant of Brouwer's Theorem requires the notion of contractibility.

We say that a topological space X is *contractible* to a point $z \in X$ if there exists a continuous mapping *(a homotopy)* $H : X \times [0, 1] \to X$ satisfying $H(x, 0) = x$ and $H(x, 1) = z$ for all $x \in X$. Obviously, if a topological space X is contractible to some point, then it is contractible to any point in X.

Theorem 9.4 (No contractibility theorem) *The sphere S^{n-1} is not contractible to a point.*

To see this, suppose that there is a homotopy $H : S^{n-1} \times [0, 1] \to S^{n-1}$ such that for all $x \in S^{n-1}$ we have $H(x, 0) = x$ and $H(x, 1) = z$, where $z \in S^{n-1}$ is arbitrarily fixed. Then, for any $0 \le r < 1$, the mapping defined as

$$Rx = \begin{cases} z & \text{if } \|x\| \le r, \\ H\left(\frac{x}{\|x\|}, \frac{1 - \|x\|}{1 - r}\right) & \text{if } \|x\| \in (r, 1], \end{cases}$$

would be a retraction of B^n onto S^{n-1}, which contradicts "No retraction theorem". To see that this theorem is equivalent to the Brouwer's Theorem, suppose that there is a retraction $R : B^n \to S^{n-1}$. Then, S^{n-1} would be contractible by a homotopy $H : S^{n-1} \times [0, 1] \to S^{n-1}$ defined as

$$H(x, t) = R((1 - t) x).$$

Notice that actually neither of the above constructions takes advantage of the fact that we work with a finite dimensional Banach space $\mathbb{R}^n$ and that in fact they allow the formulation of the following result, among specialists, referred to as a "trivial theorem".

Theorem 9.5 (trivial theorem) *For any Banach space X, the following statements are equivalent:*

a) *The unit ball B has the topological fixed point property,*

b) *The unit sphere S is not the retract of B,*

c) *S is not contractible.*

Classical Brouwer's Theorem shows that the above statements are true if $X = \mathbb{R}^n$, and, consequently, so are they for all finite-dimensional Banach spaces.

The remaining part of the chapter focuses on results showing how strongly three statements mentioned above are false in the case of infinite-dimensional Banach spaces. A crucial reason for this involves the fact that a bounded and closed set does not have to be compact in this case. In particular, this is applicable to every closed ball. The most popular and useful result extending Brouwer's theorem to infinite-dimensional spaces is the following theorem formulated in 1930 by Schauder [74]:

Theorem 9.6 (Schauder's Fixed Point Theorem) *Any convex and compact subset C of a Banach space X has topological fixed point property.*

Already at this very stage, a natural, general question arises. What happens if C is noncompact?

Around 1935, Ulam posed the following question in the *Scottish Book* [59]:

"Can one transform continuously the solid sphere of a Hilbert space into its boundary such that the transformation should be identity on the boundary of the ball?"

A note accompanying Ulam's question reads: *"There exists a transformation with the required property given by Tychonoff"*. Unfortunately, until now all attempts to find the Tychonoff's construction have failed. An answer to Ulam's question came eight years later and was formulated by Kakutani [45]. In his publication, Kakutani provided some examples of uniformly continuous and lipschitzian mappings without fixed points. Each of them could be used to construct the retraction required.

Example 9.1 (Kakutani's Construction) *Consider the space ℓ_2, the standard model of a Hilbert space. The mapping $T : B \to B$ defined for any $x = (x_1, x_2, \dots)$ by*

$$Tx = \left(\sqrt{1 - \|x\|}, x_1, x_2, \dots \right)$$

is fixed point free and uniformly continuous. Indeed, $Tx = x$ implies $x_i = \sqrt{1 - \|x\|} = 0$ for all $i \in \mathbb{N}$ and $\|x\| = 1$, which is impossible. Its continuity and even uniform continuity follows from the easily justified inequality

$$\|Tx - Ty\| \le \sqrt{2 \|x - y\|} + \|x - y\|.$$

For the mapping $T_\epsilon : B \to B$, $\epsilon \in (0, 1]$, defined by

$$T_\epsilon x = (\epsilon (1 - \|x\|), x_1, x_2, \dots)$$

we get the same conclusion, but this time the mapping T_ϵ is even lipschitzian with $k(T_\epsilon) = \sqrt{1 + \epsilon^2}$.

To construct a retraction $R : B \to S$, apply the following Kakutani's receipt. For any $x \in B$, follow the straight half line from Tx through x until you reach the sphere S. Then, define the value of the retraction Rx as the unique point of intersection of this half line with the unit sphere S. The situation is illustrated in figure 9.1.

The corresponding analytic formula is

$$Rx = x + \lambda(x) u(x),$$

where

$$u(x) = \frac{x - Tx}{\|x - Tx\|}$$

and $\lambda(x) \geq 0$ is selected to satisfy $\|Rx\| = 1$. Standard calculations, which we leave to the reader, show that

$$\lambda(x) = -(x, u(x)) + \sqrt{1 - \|x\|^2 + (x, u(x))^2}.$$

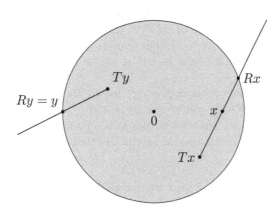

FIGURE 9.1

The next important step involves Klee's constructions formulated in 1953 and 1955 in [51] and [52]. They offered a full answer to Ulam's question not only in the case of a Hilbert space but also for any infinitely dimensional Banach space.

Theorem 9.7 (Klee's Theorem) *For any infinitely dimensional Banach space* X, *there exists a homeomorphism* $h : X \to X \setminus \{0\}$ *such that* $hx = x$ *for any* $x \in X$ *with* $\|x\| \geq 1$.

Observe that in virtue of the above result we can immediately construct the retraction $R : B \to S$ by putting

$$Rx = \frac{hx}{\|hx\|}.$$

Theorem 9.8 (Klee's Theorem) *For any noncompact, closed, and convex subset C of a Banach space X, there exists a continuous, fixed point free mapping $T : C \to C$.*

In view of Schauder's and Klee's theorems, we conclude that a closed and convex subset of a Banach space has t.f.p.p. iff it is compact.

The mappings defined by Klee are continuous but not uniformly continuous. On the other hand, there are many lipschitzian mappings known, even nonexpansive mappings without fixed points. Such observations justify the following questions.

Do all bounded, closed, convex but noncompact sets fail to have the t.f.p.p. in some stronger sense? How "regular" such mappings can be? How far they can move all the points? Do spaces or sets differ with respect to the "intensity" of this failure?

Such an approach to the problem was noticed and specified in 1973 by Goebel in [29]. It includes examples of lipschitzian mappings $T : C \to C$ for which the *minimal displacement* $d(T) = \inf_{x \in C} \|x - Tx\| > 0$. This work has exerted a major impact on the further evolution of this problem. We leave to the reader the verification of the fact that in the Kakutani's constructions we have $d(T) = 0$.

Indeed, research in subsequent years led to the formulation of theorems much stronger than Klee's construction. In 1979, Nowak [63] proved that for a certain class of Banach spaces the sphere is a lipschitzian retract of the ball. Four years later Benyamini and Sternfeld [10] proved that this is true for any Banach space.

Theorem 9.9 (Benyamini and Sternfeld) *There exists a universal constant k_0 such that any infinitely dimensional Banach space X admits a retraction $R : B_X \to S_X$ of class $L(k_0)$.*

Once we have a k-lipschitzian retraction $R : B_X \to S_X$, we can define a mapping $T = -R : B_X \to S_X \subset B_X$, which is also of class $L(k)$ and such that, for any $x \in B_X$

$$\|x - Tx\| \geq \frac{1}{k} \|Tx - T^2 x\| = \frac{2}{k} \|Rx\| = \frac{2}{k}.$$

This implies that $d(T) \geq \frac{2}{k} > 0$. Further, for any $\alpha \in (0, 1]$, the mapping $T_\alpha = (1 - \alpha) I + \alpha T \in L(1 - \alpha + \alpha k)$, $T_\alpha : B_X \to B_X$ and satisfies $d(T_\alpha) = \alpha d(T) = \frac{2\alpha}{k} > 0$. Theorem 9.9 can be also formulated as

Theorem 9.10 (Benyamini and Sternfeld) *For any infinitely dimensional Banach space X, the following three statements are true and equivalent:*

a) *For any $k > 1$, there exists a mapping $T : B_X \to B_X$ of class $L(k)$ such that $d(T) > 0$,*

b) *The unit sphere S_X is a lipschitzian retract of B_X,*

c) *S_X is Lipschitz contractible.*

The last statement means that there exist constants $M, N \geq 0$, and a homotopy $H : S_X \times [0,1] \to S_X$ such that $H(x,0) = x$ for all $x \in S$, $H(x,1) = z$ for arbitrarily fixed $z \in S_X$ and

$$\|H(x,t) - H(y,s)\| \leq M\,|t - s| + N\,\|x - y\|$$

for all $x, y \in S_X$ and for all $s, t \in [0,1]$. It is clear that, if the retraction $R : B_X \to S_X$ is of class $L(k)$, then the homotopy obtained by standard tricks, $H(x,t) = R((1-t)\,x)$, satisfies

$$\|H(x,t) - H(y,s)\| \leq k\,|t - s| + k\,\|x - y\|\,.$$

The strongest result in this matter, concerning not only ball but also all bounded, closed and convex subsets was obtained in 1985 by Lin and Sternfeld [57]:

Theorem 9.11 (Lin and Sternfeld) *Let C be a bounded, closed, convex, but noncompact subset of a Banach space X. Then, for any $k > 1$, there exists a mapping $T : C \to C$ of class $L(k)$ with $d(T) > 0$.*

In spite of the fact that theorems 9.9 and 9.11 provide qualitative answers to the theory, there are still some quantitative aspects, which were initiated by Goebel in 1973. The two basic are *the minimal displacement problem* and *optimal retraction problem*. This topic has been widely studied in books [36], [49], and [31] as well as in the survey articles [32] and [41]. In spite of this, within the past 10 years we have had big progress. The aim of this chapter is to give the updated presentation of the subject and possibly attract newcomers to the field.

To proceed, we shall need some basic facts concerning retractions of the whole space onto balls.

Exhaustive information as well as proofs of almost all above-mentioned results can be found in books [31] and [36].

9.2   Retracting onto balls in Banach spaces

The most popular retraction of the whole Banach space X onto the unit ball B_X is the radial projection $P : X \to B_X$ defined as

$$Px = \begin{cases} x & \text{if } \|x\| \leq 1, \\ \frac{x}{\|x\|} & \text{if } \|x\| > 1. \end{cases}$$

It is easy to observe that

$$\|x - Px\| = \text{dist}\,(x, B_X) = \max\{0, \|x\| - 1\}.$$

Thus, P is *the nearest point projection*. In general, P is of class $L\,(2)$ because of the following:

Lemma 9.1 *For any $x, y \in X$, $x, y \neq 0$, we have*

$$\left\| \frac{x}{\|x\|} - \frac{y}{\|y\|} \right\| \leq \frac{2}{\max\{\|x\|, \|y\|\}} \|x - y\|.$$

Proof. Observe that, for any $x, y \neq 0$,

$$\left\| \frac{x}{\|x\|} - \frac{y}{\|y\|} \right\| \leq \left\| \frac{x}{\|x\|} - \frac{x}{\|y\|} \right\| + \left\| \frac{x}{\|y\|} - \frac{y}{\|y\|} \right\|$$

$$= \|x\| \left| \frac{1}{\|x\|} - \frac{1}{\|y\|} \right| + \frac{1}{\|y\|} \|x - y\|$$

$$= \frac{1}{\|y\|} \left| \|x\| - \|y\| \right| + \frac{1}{\|y\|} \|x - y\|$$

$$\leq \frac{2}{\|y\|} \|x - y\|.$$

Exchanging roles of x and y in the above evaluations gives us the thesis.

$\square$

Example 9.2 *As a Banach space $(X, \|\cdot\|)$ consider $\ell_\infty^{(2)} = \left(\mathbb{R}^2, \|\cdot\|\right)$ furnished with the norm $\|x\| = \|(x_1, x_2)\| = \max\{|x_1|, |x_2|\}$. Then, for $x = (1,1)$ and $y_\epsilon = (1 + \epsilon, 1 - \epsilon)$, we get (see figure 9.2)*

$$\|Px - Py_\epsilon\| = \left\| (1,1) - \left(1, \frac{1-\epsilon}{1+\epsilon}\right) \right\| = \left\| \left(0, \frac{2\epsilon}{1+\epsilon}\right) \right\|$$

$$= \frac{2}{1+\epsilon} \|x - y_\epsilon\|.$$

Since $\epsilon > 0$ can be chosen arbitrarily close to 0, we get $k\,(P) = 2$.

Spaces c_0, c, ℓ_∞, $C\,[0,1]$, ℓ_1, $L_1\,(0,1)$ contain an isometric copy of $\ell_\infty^{(2)}$. Thus, $k\,(P) = 2$ for each of them. Moreover, it is known that $k\,(P) < 2$ if and only if $\epsilon_0\,(X) < 2$; see [75]. Results concerning the Lipschitz constants of radial projections in spaces ℓ_p and L_p can be found in papers [27] and [44].

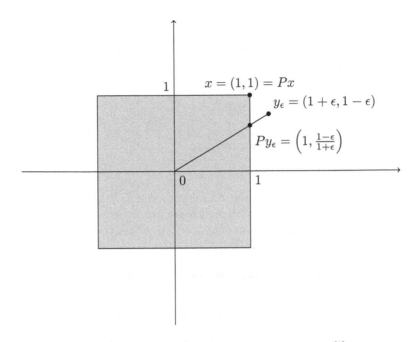

FIGURE 9.2: The radial projection in $\ell_\infty^{(2)}$.

Let us pass to the case of Hilbert space H. Suppose that C is a closed and convex subset of H. Since every Hilbert space is uniformly convex, for any point $x \in H$, there is exactly one closest point $P_C x$ in C, $\|x - P_C x\| = \inf\{\|x - z\| : z \in C\}$. Thus, the mapping $P_C : H \to C$ described above is well defined, and it is *the nearest point projection*. We shall prove that P_C is nonexpansive.

Lemma 9.2 *The nearest point projection* $P_C : H \to C$ *is of class* $L(1)$.

Proof. For any $z \in C$, consider the function $\phi : [0, 1] \to \mathbb{R}$ defined as

$$\phi(t) = \|x - (1 - t) P_C x - tz\|^2$$
$$= \|x - P_C x\|^2 + 2t (P_C x - x, z - P_C x) + t^2 \|P_C x - z\|^2.$$

Obviously, ϕ is convex and $\phi(0) \le \phi(t)$ for any $t \in [0, 1]$. This implies that

$$\phi'(0) = 2 (P_C x - x, z - P_C x) \ge 0$$

or equivalently

$$(P_C x - x, z - P_C x) \ge 0$$

for all $z \in C$. Thus, for any two points $x, y \in H$, we have

$$(P_C x - x, P_C y - P_C x) \ge 0$$

and
$$(P_C y - y, P_C x - P_C y) \geq 0.$$

This implies
$$((x - y) - (P_C x - P_C y), P_C x - P_C y) \geq 0,$$

and, consequently,
$$\|P_C x - P_C y\|^2 \leq (P_C x - P_C y, x - y)$$
$$\leq \|P_C x - P_C y\| \, \|x - y\|.$$

This shows that P_C is nonexpansive.

$\square$

Thus, for a Hilbert space, the radial projection P not only is a unique nearest point projection but also is nonexpansive.

The complete characterization of spaces, for which the radial projection is nonexpansive, has been given in [21]:

Theorem 9.12 *For the radial projection* $P : X \to B_X$ *in* $(X, \|\cdot\|)$, *the following statements hold.*

- *If* $\dim X \geq 3$, *then* P *is nonexpansive if and only if* X *is a Hilbert space.*

- *If* $\dim X = 2$, *then* P *is nonexpansive if and only if the norm* $\|\cdot\|^*$ *dual to* $\|\cdot\|$ *satisfies*
$$\|(x, y)\|^* = \|(-y, x)\|$$
for all $(x, y) \in \mathbb{R}^2$. *It means that the unit ball* B^* *coincides with the unit ball* B *rotated by* $\frac{\pi}{2}$.

Example 9.3 *Mixed* ℓ_p, ℓ_q *norms.* *Take* $p, q > 1$ *such that* $\frac{1}{p} + \frac{1}{q} = 1$. *Consider the space* $\mathbb{R}^2$ *furnished with the norm (see figure 9.3)*

$$\|(x_1, x_2)\|_{p,q} = \begin{cases} (|x_1|^p + |x_2|^p)^{\frac{1}{p}} & \text{if } x_1 \cdot x_2 \geq 0, \\ (|x_1|^q + |x_2|^q)^{\frac{1}{q}} & \text{if } x_1 \cdot x_2 \leq 0. \end{cases}$$

It is easy to verify that the norm $\|\cdot\|^*_{p,q}$, *dual to* $\|\cdot\|_{p,q}$ *is given via relation*

$$\|(y_1, y_2)\|^*_{p,q} = \max\left\{ x_1 y_1 + x_2 y_2 : \|(x_1, x_2)\|_{p,q} \leq 1 \right\}$$
$$= \begin{cases} (|y_1|^p + |y_2|^p)^{\frac{1}{p}} & \text{if } y_1 \cdot y_2 \leq 0, \\ (|y_1|^q + |y_2|^q)^{\frac{1}{q}} & \text{if } y_1 \cdot y_2 \geq 0. \end{cases}$$

Thus, $\|(x_1, x_2)\|^*_{p,q} = \|(-x_2, x_1)\|_{p,q}$ *for any* $(x_1, x_2) \in \mathbb{R}^2$. *Hence, in spite of the fact that* $\left(\mathbb{R}^2, \|\cdot\|_{p,q}\right)$ *is not a Hilbert space, the radial projection is nonexpansive.*

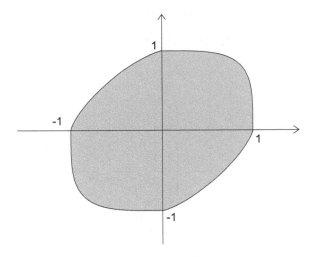

FIGURE 9.3: The unit ball in $(\mathbb{R}^2, \|\cdot\|_{p,q})$ with $p = 4$ and $q = \frac{4}{3}$.

Nevertheless, for some spaces with uniform norm, there are other than radial nonexpansive projections.

Example 9.4 *Consider again the space $\ell_\infty^{(2)}$ and the* **truncation function** *$\tau : \mathbb{R} \to [-1, 1]$ defined as (see figure 9.4)*

$$\tau(t) = \max\{-1, \min\{1, t\}\} = \begin{cases} -1 & \text{if } t < -1, \\ t & \text{if } -1 \leq t \leq 1, \\ 1 & \text{if } t > 1. \end{cases}$$

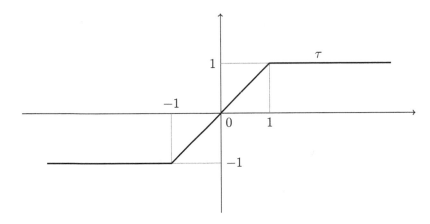

FIGURE 9.4: *A graph of function τ.*

Define $Q^{\tau} : \ell_{\infty}^{(2)} \to B$ as (see figure 9.5)

$$Q^{\tau}(x_1, x_2) = (\tau(x_1), \tau(x_2)).$$

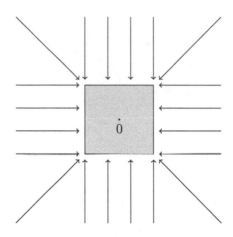

FIGURE 9.5: *Retraction $Q^{\tau} : \ell_{\infty}^{(2)} \to B$.*

It is clear that Q^{τ} is a nonexpansive retraction. Moreover, it is the nearest point projection onto B.

Example 9.5 *Consider the space $BC(\mathbb{R})$ of all bounded continuous functions $f : \mathbb{R} \to \mathbb{R}$ furnished with the standard sup norm, $\|f\| = \sup\{|f(t)| : t \in \mathbb{R}\}$. The truncation retraction $Q^{\tau} : BC(\mathbb{R}) \to B$ is given by (see figures 9.6 and 9.7)*

$$Q^{\tau} f(t) = \tau(f(t)).$$

Again, Q^{τ} is nonexpansive, and it is the nearest point projection.

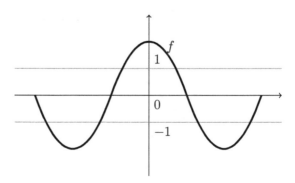

FIGURE 9.6: *A graph of function f.*

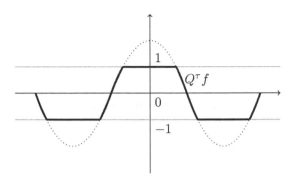

FIGURE 9.7: *A graph of $Q^\tau f$.*

Now, let us consider a more general situation. Suppose K is an arbitrary nonempty set and $B(K)$ is a Banach space consisting of all real bounded functions on K and furnished with the uniform norm, $\|f\| = \sup\{|f(x)| : x \in K\}$. A subspace X of $B(K)$ is called *truncation invariant* if for each $f \in X$ the function $Q^\tau f = \tau \circ f \in X$, where τ is the truncation. In particular, $B(K)$ is a truncation invariant subspace of $B(K)$. It is clear that the truncation $Q^\tau : X \to B_X$ is a nonexpansive retraction onto the unit ball B_X. Moreover, it is the nearest point projection. The class of truncation invariant subspaces was introduced in [39].

Let us list some particular cases:

- If $K = \{1, \dots, n\}$, then $B(K) = \ell_\infty^{(n)} = (\mathbb{R}^n, \|\cdot\|)$, where

$$\|(x_1, \dots, x_n)\| = \max\{|x_1|, \dots, |x_n|\}.$$

- If $K = \mathbb{N}$, then c, c_0, and $\ell_\infty^{(n)}$ are truncation invariant subspaces of $B(K) = \ell_\infty$.

- If $K = \mathbb{R}$ and as a norm in $\mathbb{R}$ we take an absolute value $|\cdot|$, then $BC(\mathbb{R})$ is a truncation invariant subspace of $B(\mathbb{R})$. Also, the space $BC_0(\mathbb{R})$ of all bounded continuous functions $f : \mathbb{R} \to \mathbb{R}$ vanishing at 0, $f(0) = 0$, is a truncation invariant subspace of $B(\mathbb{R})$.

- If $(K, \|\cdot\|) = ([0, 1], |\cdot|)$, then $C[0, 1]$ and $C_0[0, 1]$ are truncation invariant subspaces of $B([0, 1])$.

- Suppose (M, d) is a metric space. By $BC(M)$ we denote the space of all bounded continuous functions $f : M \to M$ and by $BC_z(M)$ its subspace, consisting of of all functions vanishing at z, $f(z) = 0$. It is clear that $BC(M)$ and $BC_z(M)$ are truncation invariant subspaces of $B(M)$.

For each $r \geq 0$, the mapping Q^τ generates the truncation retraction Q_r^τ : $X \to rB_X$ of the whole space X onto a ball centered at origin and radius r,

$$Q_r^\tau f = \begin{cases} rQ^\tau \left(\frac{1}{r} f \right) & \text{if } r > 0, \\ 0 & \text{if } r = 0. \end{cases}$$

Obviously, for every $r \geq 0$, Q_r^τ is the nearest point projection onto rB_X. In particular, for every $f \notin rB_X$, we have $\|Q_r^\tau f\| = r$.

A very useful property of this family is given in the following

Lemma 9.3 *For all $f, g \in X$, and $r_1, r_2 \geq 0$, we have*

$$\left\| Q_{r_1}^\tau f - Q_{r_2}^\tau g \right\| \leq \max \{ \|f - g\|, |r_1 - r_2| \}.$$

Proof. We shall prove that the inequality

$$\left| Q_{r_1}^\tau f(x) - Q_{r_2}^\tau g(x) \right| \leq \max \{ |f(x) - g(x)|, |r_1 - r_2| \}$$

holds for every $x \in K$.

It is clear for $r_1 = r_2 = 0$. If $r_2 = 0$ and $r_1 > 0$, then, for any $x \in K$, we have

$$\left| Q_{r_1}^\tau f(x) - Q_0^\tau g(x) \right| \leq |r_1| = |r_1 - r_2|.$$

Suppose that $0 < r_2 \leq r_1$. Let $x \in K$ be arbitrarily chosen. Then, we have the following cases:

1. If $|f(x)| \leq r_1$ and $|g(x)| \leq r_2$, then

$$\left| Q_{r_1}^\tau f(x) - Q_{r_2}^\tau g(x) \right| = \left| r_1 \tau \left(\frac{1}{r_1} f(x) \right) - r_2 \tau \left(\frac{1}{r_2} g(x) \right) \right|$$
$$= |f(x) - g(x)|.$$

2. If $f(x) \geq r_1$ and $g(x) \geq r_2$ or $f(x) \leq -r_1$ and $g(x) \leq -r_2$, then

$$\left| Q_{r_1}^\tau f(x) - Q_{r_2}^\tau g(x) \right| = |r_1 - r_2|.$$

3. If $f(x) \geq r_1$ and $|g(x)| \leq r_2$, then

$$\left| Q_{r_1}^\tau f(x) - Q_{r_2}^\tau g(x) \right| = |r_1 - g(x)| = r_1 - g(x) \leq f(x) - g(x)$$
$$= |f(x) - g(x)|.$$

A similar situation applies to $f(x) \leq -r_1$ and $|g(x)| \leq r_2$; that is,

$$\left| Q_{r_1}^\tau f(x) - Q_{r_2}^\tau g(x) \right| \leq |f(x) - g(x)|.$$

4. If $f(x) \geq r_1$ and $g(x) \leq -r_2$, then

$$\left| Q_{r_1}^\tau f(x) - Q_{r_2}^\tau g(x) \right| = |r_1 + r_2| = r_1 + r_2 \leq f(x) - g(x)$$
$$= |f(x) - g(x)|.$$

Obviously, for $f(x) \leq -r_1$ and $g(x) \geq r_2$ we also have

$$\left| Q_{r_1}^\tau f(x) - Q_{r_2}^\tau g(x) \right| \leq |f(x) - g(x)|.$$

5. If $r_2 \leq f(x) \leq r_1$ and $g(x) \geq r_2$, then

$$\left| Q_{r_1}^\tau f(x) - Q_{r_2}^\tau g(x) \right| - |f(x) - r_2| = f(x) - r_2 \leq r_1 - r_2$$
$$= |r_1 - r_2|.$$

By analogy, if $-r_1 \leq f(x) \leq -r_2$ and $g(x) \leq -r_2$, then

$$\left| Q_{r_1}^\tau f(x) - Q_{r_2}^\tau g(x) \right| \leq |r_1 - r_2|.$$

6. If $-r_1 \leq f(x) \leq r_2$ and $g(x) \geq r_2$, then

$$\left| Q_{r_1}^\tau f(x) - Q_{r_2}^\tau g(x) \right| = |f(x) - r_2| = -f(x) + r_2 \leq -f(x) + g(x)$$
$$= |f(x) - g(x)|.$$

If $-r_2 \leq f(x) \leq r_1$ and $g(x) \leq -r_2$, then we also have

$$\left| Q_{r_1}^\tau f(x) - Q_{r_2}^\tau g(x) \right| \leq |f(x) - g(x)|.$$

If $r_1 \leq r_2$, then it is enough to repeat the above cases exchanging roles between f and g. Hence, for all $r_1, r_2 \geq 0$, $f, g \in X$ and every $x \in K$, we obtain

$$\left| Q_{r_1}^\tau f(x) - Q_{r_2}^\tau g(x) \right| \leq \max \left\{ |f(x) - g(x)|, |r_1 - r_2| \right\},$$

as we desired. Consequently, the inequality

$$\left| Q_{r_1}^\tau f(x) - Q_{r_2}^\tau g(x) \right| \leq \max \left\{ \|f - g\|, |r_1 - r_2| \right\}$$

holds for every $x \in K$. By taking supremum over all $x \in K$ on the left-hand side of the above inequality, we finally get the thesis.

$\square$

For spaces like $B(M)$, $BC(M)$, and $BC_z(M)$, there are other non-expansive retractions onto the unit ball. To see it, consider any function $\beta : \mathbb{R} \to [-1, 1]$ satisfying $\beta(t) = t$ for $t \in [-1, 1]$ and $|\beta(s) - \beta(t)| \leq |s - t|$ for all $s, t \in \mathbb{R}$. Every such function generates a nonexpansive retraction $Q^\beta f = \beta \circ f$ onto the unit ball.

Consider, for example, the retraction Q^β corresponding to the function β given in figure 9.8.

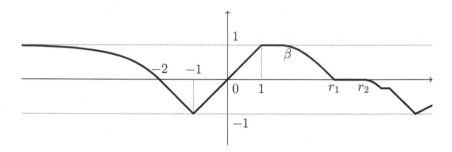

FIGURE 9.8: A graph of function β.

Such retraction maps the ball centered at $z \equiv -\frac{3}{2}$ and radius $\frac{1}{2}$ onto the ball centered at $\alpha \circ z \equiv -\frac{1}{2}$ and radius $\frac{1}{2}$. More precisely, every function $f \in B\left(z, \frac{1}{2}\right)$ is reflected with respect to the level -1, $Q^\beta f(t) = -f(t) - 2$ for $t \in M$. Observe also that the ball centered at $z \equiv \frac{r_1 + r_2}{2}$ and radius $\frac{r_2 - r_1}{2}$ is mapped to zero.

Of particular interest for us will be the retraction $Q^\Lambda f = \Lambda \circ f$ corresponding to the function $\Lambda : \mathbb{R} \to [-1, 1]$ defined as (see figure 9.9)

$$\Lambda(t) = \begin{cases} 0 & \text{if } t < -2, \\ -t - 2 & \text{if } -2 \leq t < -1, \\ t & \text{if } -1 \leq t \leq 1, \\ -t + 2 & \text{if } 1 < t \leq 2, \\ 0 & \text{if } t > 2. \end{cases}$$

To see how Q^Λ works, consider, for instance, the space $BC(\mathbb{R})$. Take any function $f \in BC(\mathbb{R})$ and divide its graph into five groups of pieces in the following way: the pieces that lie above the level 2 (resp. below the level -2), between levels 1 and 2 (resp. between levels -2 and -1) and between levels -1 and 1. Obviously, the part of graph that lies between levels -1 and 1 remains unchanged. The part which lies between levels 1 and 2 (resp. between levels -2 and -1) is reflected with respect to the level 1 (resp. level -1). The pieces lying above the level 2 as well as below the level -2 are mapped to zero. The situation is illustrated in figures 9.10 and 9.11.

As opposed to Q^τ, the retraction Q^Λ does not send all the elements to the closest point in B. Even more, for any $f \in B(M)$ with $\|f\| > 1$, we have

$$\|f - Q^\tau f\| = \operatorname{dist}(f, B) = \|f\| - 1 < \|f - Q^\Lambda f\| = \min\{2\|f\| - 2, \|f\|\}.$$

The same is true for subspaces $BC(M)$ and $BC_z(M)$.

We shall use the retractions Q^τ and Q^Λ discussing the so-called problems of *minimal displacement* and *optimal retractions*.

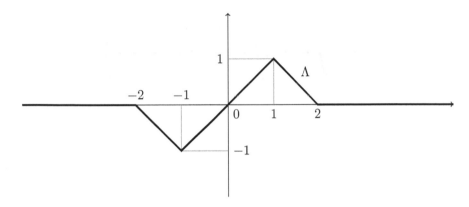

FIGURE 9.9: A graph of function Λ.

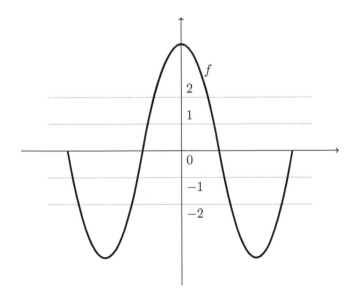

FIGURE 9.10: A graph of function f.

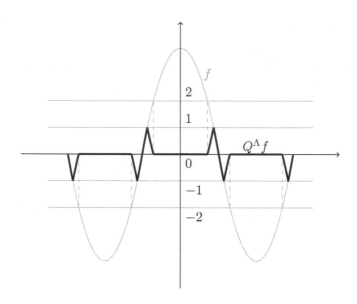

FIGURE 9.11: A graph of function $Q^\Lambda f$.

9.3   Minimal displacement

Suppose that C is a bounded, closed, convex, and noncompact subset of a Banach space X. Recall that the minimal displacement of a mapping $T : C \to C$ is the number

$$d(T) = \inf \{\|x - Tx\| : x \in C\}.$$

For any set C, the *characteristic of minimal displacement* is the function $\varphi_C : [1, +\infty) \to [0, \operatorname{diam}(C)]$ defined as

$$\varphi_C(k) = \sup \{d(T) : T : C \to C, T \in L(k)\}.$$

In this setting, theorem 9.11 states that $\varphi_C(k) > 0$ for $k > 1$.

Recall that the *Chebyshev radius* of C relative to C is the number

$$\mathrm{r}(C) = \inf_{z \in C} \sup \{\|z - y\| : y \in C\}.$$

Observe that for any two sets K and C such that $K = aC + b$, $a \neq 0$,

$b \in X$, we have $r(K) = |a| \, r(C)$. Further, for every mapping $T : C \to C$ of class $L(k)$, we can define the mapping $T_1 : K \to K$ as

$$T_1 x = aT \left(\frac{x - b}{a} \right) + b.$$

Then, $k(T_1) = k(T)$ and $d(T_1) = |a| \, d(T)$. Consequently,

$$\varphi_K(k) = \varphi_{aC+b}(k) = |a| \, \varphi_C(k).$$

Thus, without loss of generality, we can always assume that $r(C) = 1$. In the special case, when $C = B_X$, we write ψ_X instead of φ_{B_X}.

For the whole Banach space X, we define the characteristic of the whole space $\varphi_X : [1, +\infty) \to [0, \mathrm{diam}(C)]$ as

$$\varphi_X(k) = \sup \{ \varphi_C(k) : C \subset X, r(C) = 1 \}.$$

As usual, we shall drop the subscripts when either C or X is clear from the context.

Let us list some general properties of functions φ_C, φ_X, and ψ_X (see [36], pp. 215–216 and 222).

Lemma 9.4 *If $r(C) = 1$ and φ denotes any of the functions φ_X, φ_C, ψ_X, then:*

(1) $\varphi(k)$ *increases with k;*

(2) $\varphi(1 - \alpha + \alpha k) \geq \alpha \varphi(k)$ *for all $\alpha \in [0,1]$ and $k > 1$;*

(3) *the ratio $\frac{\varphi(k)}{k-1}$ decreases with k;*

(4) *the ratio $\frac{k\varphi(k)}{k-1}$ increases with k;*

(5) *the derivative $\varphi'(1) := \lim\limits_{k \to 1+} \frac{\varphi(k)}{k-1}$ always exists and $\varphi'(1) > 0$;*

(6) $\varphi'(1) \left(1 - \frac{1}{k} \right) \leq \varphi(k) \leq 1 - \frac{1}{k}$ *for all $k > 1$;*

(7) $\varphi(k) = 1 - \frac{1}{k}$ *if and only if $\varphi'(1) = 1$;*

(8) $\lim\limits_{k \to \infty} \psi_X(k) = 1$ *for any space X.*

Proof. (1) is obvious. (2) follows from the observation that for any $T : C \to C$, $T \in L(k)$ and any $\alpha \in [0,1]$ the mapping $T_\alpha : C \to C$ defined as

$$T_\alpha x = (1 - \alpha) x + \alpha T x$$

is of class $L(1 - \alpha + \alpha k)$ and such that $\|x - T_\alpha x\| = \alpha \|x - Tx\|$ for any $x \in C$.

(3) follows from (2).

To prove (4), for a fixed $A > k$ and $x \in C$ consider the equation

$$y = \left(1 - \frac{1}{A}\right)x + \frac{1}{A}Ty.$$

Then, the right-hand side defines a contraction and, consequently, the above equation has a unique solution which we shall denote by Fx,

$$Fx = \left(1 - \frac{1}{A}\right)x + \frac{1}{A}TFx.$$

Then, we have

$$\|Fx - Fy\| \leq \left(1 - \frac{1}{A}\right)\|x - y\| + \frac{1}{A}\|TFx - TFy\|$$

$$\leq \left(1 - \frac{1}{A}\right)\|x - y\| + \frac{k}{A}\|Fx - Fy\|.$$

Thus,

$$F \in L\left(\frac{A-1}{A-k}\right)$$

and

$$TF \in L\left(k\frac{A-1}{A-k}\right).$$

Observe that for any $x \in C$

$$\|x - TFx\| = \frac{A}{A-1}\|Fx - TFx\| \geq \frac{Ad(T)}{A-1}.$$

Using the introduced notation, we can write

$$\varphi\left(k\left(\frac{A-1}{A-k}\right)\right) \geq \frac{A}{A-1}\varphi(k).$$

Finally, if we put $l = \frac{k(A-1)}{A-k} > k$, then the above formula takes the form

$$\frac{\varphi(l)\,l}{l-1} \geq \frac{\varphi(k)\,k}{k-1}.$$

(5) follows from (3) and (4).

To prove (6) observe that the first inequality $\varphi'(1)\left(1 - \frac{1}{k}\right) \leq \varphi(k)$ follows from (4) and (5). The second estimate $\varphi(k) \leq 1 - \frac{1}{k}$ follows from the subsequent observation. Suppose $T : C \to C$ is of class $L(k)$. Fix $\epsilon > 0$. Then, there exists $z \in C$ such that $C \subset B(z, 1 + \epsilon)$. Consider the equation

$$x = \left(1 - \frac{1}{k+\epsilon}\right)z + \frac{1}{k+\epsilon}Tx.$$

It is clear that the right-hand side defines a contraction. Thus, there is a unique point $x_\epsilon \in C$ satisfying the above equation. Then, we have

$$\|x_\epsilon - Tx_\epsilon\| = \left(1 - \frac{1}{k+\epsilon}\right)\|z - Tx_\epsilon\| \leq \left(1 - \frac{1}{k+\epsilon}\right)(1+\epsilon).$$

Letting $\epsilon \to 0$ we get $d(T) \leq 1 - \frac{1}{k}$.

(7) follows from (6).

Let us pass now to (8). In view of theorem 9.9 by Benyamini and Sternfeld, there exists the retraction $R : B_X \to S_X$ of class $L(k_0)$. Fix $\epsilon \in (0,1)$. Define the mapping $T_\epsilon : B_X \to B_X$ as

$$T_\epsilon x = \begin{cases} -R\left(\frac{x}{\epsilon}\right) & \text{if} \quad \|x\| \leq \epsilon, \\ -\frac{x}{\|x\|} & \text{if} \quad \|x\| \in [\epsilon, 1]. \end{cases}$$

Then, $T_\epsilon \in L\left(\frac{k_0}{\epsilon}\right)$. Indeed, if $x, y \in \epsilon B_X$, then

$$\|T_\epsilon x - T_\epsilon y\| = \left\|R\left(\frac{x}{\epsilon}\right) - R\left(\frac{x}{\epsilon}\right)\right\| \leq \frac{k_0}{\epsilon}\|x - y\|.$$

In view of lemma 9.1, for any $x, y \in \overline{B_X \backslash \epsilon B_X}$, we have

$$\|T_\epsilon x - T_\epsilon y\| = \left\|\frac{x}{\|x\|} - \frac{y}{\|y\|}\right\| \leq \frac{2}{\max\{\|x\|, \|y\|\}}\|x - y\|$$

$$\leq \frac{2}{\epsilon}\|x - y\| \leq \frac{k_0}{\epsilon}\|x - y\|.$$

If $\|x\| < \epsilon$ and $\|y\| > \epsilon$, then there is a unique number $\lambda \in (0,1)$ such that

$$\|(1 - \lambda)x + \lambda y\| = \epsilon.$$

Then,

$$\|T_\epsilon x - T_\epsilon y\| \leq \|T_\epsilon x - T_\epsilon((1-\lambda)x + \lambda y)\| + \|T_\epsilon((1-\lambda)x + \lambda y) - T_\epsilon y\|$$

$$\leq \lambda\frac{k_0}{\epsilon}\|x - y\| + (1-\lambda)\frac{k_0}{\epsilon}\|x - y\|$$

$$= \frac{k_0}{\epsilon}\|x - y\|,$$

as we desired.

For any $x \in B_X$, we have $\|x - T_\epsilon x\| \geq 1 - \epsilon$. Indeed, if $\|x\| \leq \epsilon$, then

$$\|x - T_\epsilon x\| = \left\|x + R\left(\frac{x}{\epsilon}\right)\right\| \geq \left\|R\left(\frac{x}{\epsilon}\right)\right\| - \|x\| \geq 1 - \epsilon,$$

and if $\|x\| \geq \epsilon$, then

$$\|x - T_\epsilon x\| = \left\|x + \frac{x}{\|x\|}\right\| = 1 + \|x\| \geq 1 + \epsilon.$$

By putting $k = \frac{k_0}{\epsilon}$ we finally get

$$\psi_X(k) \geq 1 - \frac{k_0}{k}$$

for any $k > k_0$. Now, it is enough to use (6) to finish the proof.

$\square$

Observe that (1), (2), and (5) from the above lemma imply the continuity of φ_C, φ_X and ψ_X on $[1, \infty)$.

A set C is called *extremal* if $\varphi_C(k) = 1 - \frac{1}{k}$ for all $k > 1$. A space X for which the unit ball B_X is extremal, $\psi_X(k) = 1 - \frac{1}{k}$ for all $k > 1$, is referred to as an *extremal space*.

The characteristic of minimal displacement can be considered not only for the whole class of lipschitzian mappings but also for various subclasses. The most interesting are

$$\psi_{B \to S}(k) = \sup\{d(T) : T : B \to S, T \in L(k)\}$$

and

$$\psi_{S \to \{0\}}(k) = \sup\{d(T) : T : B \to B, T \in L(k) \text{ and } T(S) = \{0\}\}.$$

The minimal displacement problem involves finding or evaluating the functions φ and ψ for concrete sets or spaces.

Let us begin by listing some examples of extremal spaces.

Example 9.6 ([36]) *Consider the space $C[0, 1]$ with the standard sup norm. Take any strictly increasing function $g \in C[0, 1]$ satisfying $g(0) = -2$ and $g(1) = 2$. Then, there is a unique point $s \in (0, 1)$ such that $g(s) = 0$. Fix $k > 1$ and consider the mapping $T_k : B \to B$ defined as*

$$T_k f = Q^\tau(k(f + g)).$$

It is clear that T_k is of class $L(k)$. We shall prove that $\|f - T_k f\| > 1 - \frac{1}{k}$ for any $f \in C[0, 1]$.

First, assume that $f(s) \leq 0$. Since $f(1) + g(1) = f(1) + 2 \geq 1$, there exists a point $t_1 \in (s, 1)$ such that $f(t_1) + g(t_1) = \frac{1}{k}$. Thus,

$$\|T_k f - f\| \geq |T_k f(t_1) - f(t_1)| = |1 - f(t_1)| = \left|1 - \frac{1}{k} + g(t_1)\right|$$

$$= 1 - \frac{1}{k} + g(t_1) > 1 - \frac{1}{k}.$$

Now, assume that $f(s) \geq 0$. Since $f(0) + g(0) = f(0) - 2 \leq -1$, there is a point $t_2 \in (0, s)$ satisfying

$$f(t_2) + g(t_2) = -\frac{1}{k}.$$

Then,

$$\|T_k f - f\| \geq |T_k f(t_2) - f(t_2)| = |-1 - f(t_2)| = \left| -1 + \frac{1}{k} + g(t_2) \right|$$

$$= 1 - \frac{1}{k} - g(t_2) > 1 - \frac{1}{k}$$

as we desired. Consequently, $d(T_k) = 1 - \frac{1}{k}$. However, T_k maps B into the unit sphere S. Thus, we can write

$$\varphi_{C[0,1]}(k) = \psi_{C[0,1]}(k) = \psi_{B \to S}(k) = 1 - \frac{1}{k}.$$

Example 9.7 *Now, consider the subspace $C_0[0,1]$ of the space $C[0,1]$. Take any strictly increasing function $g \in C_0[0,1]$ satisfying $g(1) = 1$. For fixed $k > 1$ define the mapping $T_k : B \to B$ as*

$$T_k f = Q^\tau (k(|f| + g)).$$

Obviously, T_k is of class $L(k)$. Further, observe that for any $f \in B$ we have $|f(0)| + g(0) = 0$ and $|f(1)| + g(1) \geq 1$. Thus, for any $f \in B$, there exists $s \in (0,1)$ satisfying $|f(s)| + g(s) = \frac{1}{k}$ and in a similar way as in example 9.6 we get $\|f - T_k f\| > 1 - \frac{1}{k}$. Observe also that T_k maps the whole unit ball B into the unit sphere S and more precisely into its positive part $S^+ = \{f \in S : f(t) \geq 0 \text{ for all } t \in [0,1]\}$. Thus,

$$\varphi_{C_0[0,1]}(k) = \psi_{C_0[0,1]}(k) = \psi_{B \to S}(k) = \psi_{B \to S^+}(k) = 1 - \frac{1}{k}.$$

Example 9.8 *Consider the space $BC(\mathbb{R})$ furnished with the sup norm. Take any strictly increasing function $g \in BC(\mathbb{R})$ satisfying $\lim_{t \to \pm\infty} g(t) = \pm 3$ and $g(0) = 0$. For fixed $k > 1$ define mapping $T_k : B \to S$ by putting for each $f \in B$*

$$T_k f = Q^\tau (k(f + g)).$$

Again, T_k is of class $L(k)$ and $\|f - T_k f\| > 1 - \frac{1}{k}$ for any $f \in BC(\mathbb{R})$. We get it in a way known from example 9.6. Thus,

$$\varphi_{BC(\mathbb{R})}(k) = \psi_{BC(\mathbb{R})}(k) = \psi_{B \to S}(k) = 1 - \frac{1}{k}.$$

Observe also that the subspace $BC_0(\mathbb{R})$ is invariant under T_k. This implies that for any $k > 1$

$$\varphi_{BC_0(\mathbb{R})}(k) = \psi_{BC_0(\mathbb{R})}(k) = \psi_{B \to S}(k) = 1 - \frac{1}{k}.$$

Example 9.9 ([69]) *Consider the space c furnished with the sup norm.*

For fixed $k > 1$ define the mapping $T_k : B \to S$ by putting for each $x = (x_1, x_2, \ldots) \in B$

$$T_k x = Q^\tau (1, -1, kx_1, kx_2, \ldots) = (1, -1, \tau(kx_1), \tau(kx_2), \ldots).$$

It is easy to check that T_k is of class $L(k)$. We shall prove that for any $x \in B$ we have $\|T_k x - x\| > 1 - \frac{1}{k}$. Indeed, let us assume that there exists $x \in B$ such that $\|T_k x - x\| \leq 1 - \frac{1}{k}$. However, this implies that $x_{2i-1} \geq \frac{1}{k}$ and $x_{2i} \leq -\frac{1}{k}$ for $i = 1, 2, \ldots$, which is impossible because $x \in c$. Thus, $d(T_k) = 1 - \frac{1}{k}$ and

$$\varphi_c(k) = \psi_c(k) = \psi_{B \to S}(k) = 1 - \frac{1}{k}.$$

One can observe that the subspace c_0 of c is invariant under T_k. Consequently, for any $k > 1$ we get

$$\varphi_{c_0}(k) = \psi_{c_0}(k) = \psi_{B \to S}(k) = 1 - \frac{1}{k}.$$

In the above examples, we have $\varphi(k) = \psi(k) = 1 - \frac{1}{k}$ for $k > 1$. The same is true for all subspaces of $C[0, 1]$ of finite codimension (see [12]). It is still unknown if the space ℓ_∞ furnished with the sup norm has extremal balls. Very recently Bolibok [14] has proved that

$$\psi_{l_\infty}(k) \geq \begin{cases} (3 - 2\sqrt{2})(k - 1) & \text{for } 1 \leq k \leq 2 + \sqrt{2}, \\ 1 - \frac{2}{k} & \text{for } k > 2 + \sqrt{2}. \end{cases}$$

The same estimate holds for $L_1(0, 1)$ equipped with the standard norm (see [11]) as well as for few other spaces (see [39]). It is known that for any space X with the characteristic of convexity $\epsilon_0(X) < 1$ we have $\varphi_X'(1) < 1$ and $\varphi_X(k) < 1 - \frac{1}{k}$ for all $k > 1$ (for the proof see [36], pp. 217–218). In particular, it applies to all uniformly convex spaces.

An interesting situation is observed for the space ℓ_1. It is known that this space is not extremal, and we have (for the proof see [36], pp. 212, ex. 20.3 or [31], pp. 145, ex. 12.5)

$$\psi_{\ell_1}(k) \leq \begin{cases} \frac{2+\sqrt{3}}{4}\left(1 - \frac{1}{k}\right) & \text{for } 1 \leq k \leq 3 + 2\sqrt{3}, \\ \frac{k+1}{k+3} & \text{for } k > 3 + 2\sqrt{3}. \end{cases} \tag{9.1}$$

Nevertheless, ℓ_1 contains an extremal subset as seen in the following:

Example 9.10 ([31]) *Consider the "positive face" S^+ of the unit sphere in the space ℓ_1:*

$$S^+ = \left\{ \{x_i\}_{i=1}^\infty : x_i \geq 0, \sum_{i=1}^\infty x_i = 1 \right\}.$$

Obviously, $r(S^+) = diam(S^+) = 2$. For fixed $k > 1$, define the map $i : S^+ \to \mathbb{N}$ by

$$i(x) = \max\left\{ i \in \mathbb{N} : \sum_{j=i}^{\infty} x_i > \frac{1}{k} \right\}.$$

Let $\mu(x) \in [0,1)$ be given via relation

$$\mu(x)x_{i(x)} + \sum_{j=i(x)+1}^{\infty} x_j = \frac{1}{k}.$$

Then the mapping $T : S^+ \to S^+$ defined by

$$Tx = T(x_1, x_2, \ldots) = k\left(0, 0, \ldots, 0, \mu(x)x_{i(x)}, x_{i(x)+1}, x_{i(x)+2}, \ldots\right),$$

where 0 in the last sequence appears $i(x)$ times, is of class $L(k)$ and

$$d(T) = 2\left(1 - \frac{1}{k}\right).$$

Indeed, let $x, y \in S^+$. We shall consider the case in which $i(x) \neq i(y)$. The case $i(x) = i(y)$ is similar. Without loss of generality, we can suppose that $i(x) < i(y)$ and then $\|Tx - Ty\|$ is equal to

$$k\left(\mu(x)x_{i(x)} + \sum_{j=i(x)+1}^{i(y)-1} x_j + \left|x_{i(y)} - \mu(y)y_{i(y)}\right| + \sum_{j=i(y)+1}^{\infty} |x_j - y_j|\right)$$

$$\leq k\left(\frac{1}{k} - \sum_{j=i(x)+1}^{\infty} x_j + \sum_{j=i(x)+1}^{i(y)-1} x_j + \left|x_{i(y)} - y_{i(y)}\right| + (1 - \mu(y))\,y_{i(y)}\right.$$

$$\left. + \sum_{j=i(y)+1}^{\infty} |x_j - y_j|\right)$$

$$\leq k\left(\frac{1}{k} - \sum_{j=i(y)}^{\infty} x_j - \frac{1}{k} + \sum_{j=i(y)}^{\infty} y_j + \sum_{j=i(y)}^{\infty} |x_j - y_j|\right)$$

$$= k\left(\sum_{j=1}^{i(y)-1} x_j - \sum_{j=1}^{i(y)-1} y_j + \sum_{j=i(y)}^{\infty} |x_j - y_j|\right)$$

$$\leq k\sum_{j=1}^{\infty} |x_j - y_j|$$

$$= k\,\|x - y\|.$$

For every $x = (x_1, x_2, \ldots) \in S^+$, we have

$$\|x - Tx\| = \sum_{j=1}^{i(x)} x_j + \left|x_{i(x)+1} - k\mu(x)x_{i(x)}\right| + \sum_{j=i(x)+1}^{\infty} |x_{j+1} - kx_j|$$

$$\geq 1 - \frac{1}{k} + \mu(x)x_{i(x)} - x_{i(x)+1} + k\mu(x)x_{i(x)} + k \sum_{j=i(x)+1}^{\infty} x_j$$

$$- \sum_{j=i(x)+1}^{\infty} x_{j+1}$$

$$= 1 - \frac{1}{k} + 1 + \mu(x)x_{i(x)} - \sum_{j=i(x)+1}^{\infty} x_j$$

$$\geq 2\left(1 - \frac{1}{k}\right),$$

as we desired. Hence, $\varphi_{S^+}(k) = \left(1 - \frac{1}{k}\right) r(S^+)$ *for all* $k > 1$ *and*

$$\psi_{\ell_1}(k) < \varphi_{\ell_1}(k) = 1 - \frac{1}{k}.$$

A similar construction can be done for $L_1(0,1)$.

Example 9.11 ([31]) *Let* S^+ *be the "positive face" of the unit ball in* $L_1(0,1)$; *that is,*

$$S^+ = \left\{ f \in L_1(0,1) : f \geq 0 \text{ and } \int_0^1 f(t)dt = 1 \right\}.$$

Then, $r(S^+) = diam(S^+) = 2$. *Fix* $k > 1$. *For any function* $f \in S^+$, *we define the number* t_f *via relation*

$$t_f = \sup\left\{ t : \int_0^t f(s)ds = 1 - \frac{1}{k} \right\}.$$

The reader can check that the mapping $T : S^+ \to S^+$ *given by*

$$Tf(t) = \begin{cases} 0 & \text{if } t \leq t_f, \\ kf(t) & \text{if } t > t_f, \end{cases}$$

is k-*lipschitzian and for every* $f \in S^+$,

$$\|f - Tf\| = 2\left(1 - \frac{1}{k}\right) = \left(1 - \frac{1}{k}\right) r(S^+).$$

Consequently, $d(T) = 2\left(1 - \frac{1}{k}\right)$ *and* $\varphi_{S^+}(k) = 2\left(1 - \frac{1}{k}\right)$. *This implies that, for every* $k > 1$, $\varphi_{L_1(0,1)}(k) = 1 - \frac{1}{k}$.

We have already seen that, for the space ℓ_1, we have

$$\psi_{\ell_1}(k) < \varphi_{\frac{1}{2}S^+}(k) = 1 - \frac{1}{k}.$$

In the case of Hilbert space H, the situation is different, and for any set $C \subset H$ with a Chebyshev radius equal to 1, we have

$$\varphi_C(k) = \psi_H(k) \quad \text{for all } k > 1.$$

Indeed, suppose that $T : C \to C$ is of class $L(k)$. Then, an extension $\tilde{T} : H \to C$ of T given by

$$\tilde{T}x = TPx,$$

where P denotes the nearest point projection onto C (see lemma 9.2), is also k-lipschitzian. Moreover, by nonexpansiveness of P, we get

$$\left\| x - \tilde{T}x \right\| = \| x - TPx \| \geq \| Px - PTPx \| = \| Px - TPx \|.$$

Consequently, $d\left(\tilde{T}\right) = d(T)$. Now, it is enough to note that there is a unique point $z \in H$ such that $C \subset B(z, 1)$ and consider the restriction of $\tilde{T}$ to this ball.

For a Hilbert space H, we have (for the proof see [29] or [31], pp. 167, or [36], pp. 214, ex. 20.4):

$$\varphi_H(k) = \psi_H(k) \leq \left(1 - \frac{1}{k}\right)\left(\frac{k}{k+1}\right)^{\frac{1}{2}} \quad \text{and} \quad 0 < \psi_H'(k) \leq \frac{1}{\sqrt{2}}.$$

It is worth stressing that the above estimate was obtained 40 years ago, and so far it is not known whether it is accurate.

As opposed to the spaces $C[0, 1]$, $BC(\mathbb{R})$, $C_0[0, 1]$, $BC_0(\mathbb{R})$, c and c_0, in the case of Hilbert space both functions, ψ_H and $\psi_{B \to S}$, differ at least in the vicinity of 1. To see this, consider the following example.

Example 9.12 ([31]) *Consider the Hilbert space H and let $T : B \to S$ be a mapping of class $L(k)$. Then, the mapping $\frac{1}{k}T$ is nonexpansive and maps B into the sphere $S\left(\frac{1}{k}\right) \subset B$. Since every Hilbert space has the fixed point property for nonexpansive mappings, there exists a point $x \in B$ satisfying $x = \frac{1}{k}Tx$. Obviously, $\|x\| = \frac{1}{k}$. For the same reason, there is a point $y \in B$, satisfying*

$$y = \left(1 - \frac{1}{k}\right)x + \frac{1}{k}Ty.$$

Then,

$$k^2 \|x - y\|^2 \geq \|Tx - Ty\|^2 = \|kx - ky + (k-1)x\|^2$$
$$= k^2 \|x - y\|^2 + 2k(k-1)(x - y, x) + (k-1)^2 \|x\|^2$$

and, consequently,

$$2k(k-1)(y - x, x) \geq \frac{(k-1)^2}{k^2}.$$

Further,

$$1 = \|Ty\|^2 = \|(Ty - x) + x\|^2 = \|Ty - x\|^2 + 2(Ty - x, x) + \|x\|^2$$

$$= \left(\frac{k}{k-1}\right)^2 \|Ty - y\|^2 + 2k(y - x, x) + \frac{1}{k^2}$$

$$\geq \left(\frac{k}{k-1}\right)^2 d(T)^2 + \frac{k-1}{k^2} + \frac{1}{k^2}$$

$$= \left(\frac{k}{k-1}\right)^2 d(T)^2 + \frac{1}{k}.$$

Thus, $d(T) \leq \left(1 - \frac{1}{k}\right)\sqrt{1 - \frac{1}{k}}$, *and, consequently,*

$$\psi_{B \to S}(k) \leq \left(1 - \frac{1}{k}\right)^{\frac{3}{2}}. \tag{9.2}$$

Thus, $\psi'_{B \to S}(1) = 0$, *whereas* $\psi'(1) > 0$. *It is not clear whether* $\psi_{B \to S}(k) < \psi(k)$ *for all* $k > 1$.

An interesting situation is observed for the function $\psi_{S \to \{0\}}$. Let us begin with some general estimates, true for any space (see [31]).

Let X be an arbitrary Banach space. Consider any mapping $T : B \to B$ of class $L(k)$ sending all the points on the unit sphere to the origin, $T(S) = \{0\}$. Extend T to the mapping $\tilde{T}$ defined on the whole space X by putting $\tilde{T}x = 0$ if $\|x\| \geq 1$. Observe that for any $x \neq 0$, $x \in B$, we have

$$\left\| \tilde{T}x - \frac{1}{2}\tilde{T}0 \right\| \leq \frac{1}{2}\|Tx\| + \frac{1}{2}\|Tx - T0\|$$

$$\leq \frac{1}{2}\left\| Tx - T\left(\frac{x}{\|x\|}\right) \right\| + \frac{k}{2}\|x\|$$

$$\leq \frac{k}{2}(1 - \|x\|) + \frac{k}{2}\|x\| = \frac{k}{2}.$$

Also if $\|x\| \geq 1$, or $x = 0$, then

$$\left\| \tilde{T}x - \frac{1}{2}\tilde{T}0 \right\| = \frac{1}{2}\|T0\| \leq \frac{1}{2}.$$

Thus, $\tilde{T}$ maps the ball centered at $\frac{\tilde{T}0}{2} = \frac{T0}{2}$ of radius $\frac{k}{2}$ into itself. Hence,

$$d(T) = d\left(\tilde{T}\right) \leq \frac{k}{2}\psi_X(k)$$

and

$$\psi_{S \to \{0\}}(k) \leq \min\left\{\frac{k}{2}\psi(k), \psi(k)\right\} \leq \min\left\{\frac{k-1}{2}, \frac{k-1}{k}\right\} \tag{9.3}$$

$$= \begin{cases} \frac{k-1}{2} & \text{for } 1 \leq k \leq 2, \\ 1 - \frac{1}{k} & \text{for } k > 2. \end{cases}$$

Consequently,

$$\psi'_{S\to\{0\}}(1) = \limsup_{k\to 1^+} \frac{\psi_{S\to\{0\}}(k)}{k-1} \le \frac{1}{2}\psi'(1) \le \frac{1}{2}.$$

We leave to the reader the verification of the fact that for any Banach space we have $\lim_{k\to\infty}\psi_{S\to\{0\}}(k) = 1$. There exist spaces, for which $\psi'_{S\to\{0\}}(1) = \frac{1}{2}$.

Example 9.13 ([39]) *Let K be an infinite set, and suppose that X is extremal and truncation invariant subspace of $B(K)$. Fix $k > 1$ and take $\epsilon \in \left(0, 1 - \frac{1}{k}\right)$. There exists a mapping $T : B_X \to B_X$ of class $L(k)$ with $d(T) \ge 1 - \frac{1}{k} - \epsilon$. Extend T to the mapping $F : 2B_X \to B_X$ by putting*

$$Ff = \begin{cases} Tf & if \quad \|f\| \le 1, \\ TQ^\tau f & if \quad \|f\| \in \left[1, 2 - \frac{1}{k} - \epsilon\right], \\ Q^\tau_{\frac{k}{1+k\epsilon}(2-\|f\|)} TQ^\tau f & if \quad \|f\| \in \left[2 - \frac{1}{k} - \epsilon, 2\right]. \end{cases}$$

First, observe that on the boundary of each region, appropriate formulas coincide and $F(2S_X) = \{0\}$.
Again, $F \in L(k)$. Indeed,

- *if $f, g \in B_X$, then $\|Ff - Fg\| = \|Tf - Tg\| \le k\|f - g\|$;*

- *if $\|f\|, \|g\| \in \left[1, 2 - \frac{1}{k} - \epsilon\right]$, then having used the fact that Q^τ is nonexpansive, we have*

$$\|Ff - Fg\| = \|TQ^\tau f - TQ^\tau g\| \le k\|Q^\tau f - Q^\tau g\| \le k\|f - g\|;$$

- *if $\|f\|, \|g\| \in \left[2 - \frac{1}{k} - \epsilon, 2\right]$, then according to lemma 9.3, we get the estimate*

$$\begin{aligned} \|Ff - Fg\| &= \left\| Q^\tau_{\frac{k}{1+k\epsilon}(2-\|f\|)} TQ^\tau f - Q^\tau_{\frac{k}{1+k\epsilon}(2-\|g\|)} TQ^\tau g \right\| \\ &\le \max\left\{ \|TQ^\tau f - TQ^\tau g\|, \frac{k}{1+k\epsilon}\big|\|g\| - \|f\|\big| \right\} \\ &\le \max\{ k\|f - g\|, k\|f - g\| \} \\ &= k\|f - g\|. \end{aligned}$$

The above implies that for all $f, g \in 2B_X$,

$$\|Ff - Fg\| \le k\|f - g\|$$

as we desired.
Observe also that, for any $f \in 2B_X$, we have

$$\|f - Ff\| \ge 1 - \frac{1}{k} - \epsilon.$$

Indeed,

- *for $f \in B_X$ it is obvious;*

- *if $\|f\| \in \left[1, 2 - \frac{1}{k} - \epsilon\right]$, then*

$$\|f - Ff\| = \|f - TQ^\tau f\| \geq \|Q^\tau f - Q^\tau TQ^\tau f\| = \|Q^\tau f - TQ^\tau f\|$$
$$\geq 1 - \frac{1}{k} - \epsilon;$$

- *if $\|f\| \in \left[2 - \frac{1}{k} - \epsilon, 2\right]$, then*

$$\|f - Ff\| = \left\|f - Q^\tau_{\frac{k}{1+k\epsilon}(2-\|f\|)} TQ^\tau f\right\| \geq \|f\| - \left\|Q^\tau_{\frac{k}{1+k\epsilon}(2-\|f\|)} TQ^\tau f\right\|$$
$$\geq 2 - \frac{1}{k} - \epsilon - 1 = 1 - \frac{1}{k} - \epsilon.$$

Now, putting

$$\overline{F}f = \frac{1}{2}F(2f),$$

we obtain a mapping $\overline{F} : B_X \to B_X$ of class $L(k)$ such that $\overline{F}(S_X) = \{0\}$ and for any $f \in B_X$

$$\|f - \overline{F}f\| = \frac{1}{2}\|2f - F(2f)\| \geq \frac{1}{2}\left(1 - \frac{1}{k} - \epsilon\right).$$

Finally, by taking T of class $L(k)$ with $d(T)$ close to $\psi_X(k) = 1 - \frac{1}{k}$, we end up with the estimate

$$\psi_{S_X \to \{0\}}(k) \geq \frac{1}{2}\psi_X(k) = \frac{1}{2}\left(1 - \frac{1}{k}\right).$$

Consequently, having considered 9.3, we obtain $\psi'_{S_X \to \{0\}}(1) = \frac{1}{2}$.

The above estimate is highly inaccurate given the fact that, for any Banach space, we have $\lim_{k \to \infty} \psi_{S \to \{0\}}(k) = 1$. We have already seen that, for any space X, $\psi_{S \to \{0\}}(k) < \psi_X(k)$ for all $k \in (1, 2)$. In the light of foregoing, the following questions arise.

Are there spaces for which $\psi_{S \to \{0\}}(k) = \psi_X(k)$ for sufficiently large k? Moreover, is there a space X, for which $\psi_{S \to \{0\}}(k) = 1 - \frac{1}{k}$ on a certain subset of $[2, \infty)$?

The construction presented in the following example shows that answers to all the above questions are **affirmative**.

Example 9.14 ([41]) *Let us consider the space $C_0[0, 1]$ with the standard sup norm. In [38], it was proved that $\psi_{S \to \{0\}}(3) = \frac{2}{3}$. Below we present a general situation, for arbitrary $k > 1$.*

Let us consider two cases, for $k \in (1, 3)$ and for $k \geq 3$.

If $k \geq 3$, then we define the mapping $F_k : B \rightarrow B$ by

$$F_k f = \begin{cases} Q^\Lambda \left(k \left(|f| + g \right) \right) & \text{if } 0 \leq \|f\| \leq 1 - \frac{1}{k}, \\ Q^\tau_{k(1-\|f\|)} Q^\Lambda \left(k \left(|f| + g \right) \right) & \text{if } 1 - \frac{1}{k} \leq \|f\| \leq 1. \end{cases}$$

Obviously, $F_k (S) = \{0\}$ and in view of lemma 9.3 we conclude that F_k is of class $L(k)$.

We shall prove that, for any $f \in B$, $\|F_k f - f\| \geq 1 - \frac{1}{k}$. If $\|f\| \leq 1 - \frac{1}{k}$, then by applying a similar reasoning to this presented in example 9.7 we can select a point s satisfying $|f(s)| + g(s) = \frac{1}{k}$ to obtain $\|f - T_k f\| > 1 - \frac{1}{k}$. If $\|f\| \in \left[1 - \frac{1}{k}, 1 \right)$, then by taking a point $t_0 \in (0,1]$ such that $|f(t_0)| = \|f\|$ and using the fact that $\frac{2}{k} \leq 1 - \frac{1}{k}$ if and only if $k \geq 3$ we get the following estimate

$$\begin{aligned} \|f - F_k f\| &\geq \left| f(t_0) - k \left(1 - \|f\| \right) \tau \left(\frac{\Lambda \left(k \left(|f(t_0)| + g(t_0) \right) \right)}{k \left(1 - \|f\| \right)} \right) \right| \\ &= \left| f(t_0) - k \left(1 - \|f\| \right) \tau \left(\frac{0}{k \left(1 - \|f\| \right)} \right) \right| \\ &= |f(t_0)| \geq 1 - \frac{1}{k}. \end{aligned}$$

If $\|f\| = 1$, then $\|f - F_k f\| = \|f - 0\| = 1 > 1 - \frac{1}{k}$.

This implies that

$$\psi_{S \rightarrow \{0\}}(k) = \psi_{C_0[0,1]}(k) = 1 - \frac{1}{k}$$

for every $k \geq 3$.

Consider now the case $k \in (1,3)$. Define the mapping $F_k : \frac{4}{k+1} B \rightarrow \frac{4}{k+1} B$ as

$$F_k f = \begin{cases} Q^\Lambda \left(k \left(|f| + g \right) \right) & \text{if } 0 \leq \|f\| \leq \frac{3k-1}{k(k+1)}, \\ Q^\tau_{k\left(\frac{4}{k+1} - \|f\| \right)} Q^\Lambda \left(k \left(|f| + g \right) \right) & \text{if } \frac{3k-1}{k(k+1)} \leq \|f\| \leq \frac{4}{k+1}. \end{cases}$$

It is clear that $F_k \left(\frac{4}{k+1} S \right) = \{0\}$. Again, using lemma 9.3, we conclude that F_k is of class $L(k)$.

We shall prove that, for all $f \in \frac{k+1}{4} B$, we have $\|f - F_k f\| \geq 1 - \frac{1}{k}$. For $f \in \frac{4}{k+1} S$, it is clear. For $f \in \frac{3k-1}{k(k+1)} B$, we get it in a way known from example 9.7. For f with $\|f\| \in \left[\frac{3k-1}{k(k+1)}, \frac{4}{k+1} \right]$, we take $t_0 \in (0,1]$ such that $|f(t_0)| = \|f\|$ and consider two cases.

If $\|f\| \in \left[\frac{3k-1}{k(k+1)}, \frac{2}{k} \right]$, then we obtain

$$\|f - F_k f\| \geq |f(t_0)| - \left| k \left(\frac{4}{k+1} - |f(t_0)| \right) \tau \left(\frac{\Lambda \left(k \left(|f(t_0)| + g(t_0) \right) \right)}{k \left(\frac{4}{k+1} - |f(t_0)| \right)} \right) \right|$$

$$\geq |f(t_0)| - k\left(\frac{4}{k+1} - |f(t_0)|\right)\tau\left(\frac{\Lambda\left(k\,|f(t_0)|\right)}{k\left(\frac{4}{k+1} - |f(t_0)|\right)}\right)$$

$$= |f(t_0)| - k\left(\frac{4}{k+1} - |f(t_0)|\right)\frac{2 - k\,|f(t_0)|}{k\left(\frac{4}{k+1} - |f(t_0)|\right)}$$

$$= |f(t_0)| - 2 + k\,|f(t_0)| = (k+1)\,|f(t_0)| - 2$$

$$\geq (k+1)\frac{3k-1}{k(k+1)} - 2 = 1 - \frac{1}{k}.$$

If $\|f\| \in \left[\frac{2}{k}, \frac{4}{k+1}\right)$, then we get

$$\|f - F_k f\| \geq |f(t_0)| - k\left(\frac{4}{k+1} - |f(t_0)|\right)\tau\left(\frac{\Lambda\left(k\,|f(t_0)|\right)}{k\left(\frac{4}{k+1} - |f(t_0)|\right)}\right)$$

$$= |f(t_0)| - k\left(\frac{4}{k+1} - |f(t_0)|\right)\tau\left(\frac{0}{k\left(\frac{4}{k+1} - |f(t_0)|\right)}\right)$$

$$= |f(t_0)| \geq \frac{2}{k} \geq 1 - \frac{1}{k}.$$

Now, it is enough to define the mapping $\overline{F_k} : B \to B$ by putting for every $f \in B$

$$\overline{F_k}f = \frac{k+1}{4}F_k\left(\frac{4}{k+1}f\right),$$

which is of class $L\,(k)$, $\overline{F_k}(S) = \{0\}$, and

$$\|f - \overline{F_k}f\| = \frac{k+1}{4}\left\|\frac{4}{k+1}f - F_k\left(\frac{4}{k+1}f\right)\right\|$$

$$\geq \frac{k+1}{4}\left(1 - \frac{1}{k}\right).$$

Consequently, for every $k \in (1,3)$, we have

$$\psi_{S \to \{0\}}\,(k) \geq \frac{k+1}{4}\left(1 - \frac{1}{k}\right).$$

Taking into account (9.3), we end up with

$$\left(1 - \frac{1}{k}\right)\min\left\{\frac{k+1}{4}, 1\right\} \leq \psi_{S \to \{0\}}(k) \leq \left(1 - \frac{1}{k}\right)\min\left\{\frac{k}{2}, 1\right\}$$

for all $k \geq 1$. It is not known whether $\psi_{S \to \{0\}}\,(k) = \frac{k+1}{4}\left(1 - \frac{1}{k}\right)$ for $1 < k < 3$. However, we have an exact value of

$$\psi'_{S \to \{0\}}\,(1) = \lim_{k \to 1^+}\frac{\psi_{S \to \{0\}}\,(k)}{k - 1} = \frac{1}{2}\psi'\,(1) = \frac{1}{2}.$$

It is still open problem whether, for each Banach space X, $\psi_{S \to \{0\}}(k) = \psi_X(k)$ for some $k > 2$.

Let us pass now to the case of Hilbert space. A convenient tool in the construction presented below will be the following well-known theorem first formulated by Kirzbraun [50] for Euclidean space and then extended to any Hilbert space by Valentine [76].

Theorem 9.13 (Kirzbraun-Valentine Theorem) *Let A be an arbitrary subset of a Hilbert space H. Suppose that $T : A \to H$ is k-lipschitzian. Then, there exists a k-lipschitzian extension $\widetilde{T} : H \to Conv\,(T\,(A))$ of T.*

Example 9.15 ([31]) *Let H be a Hilbert space and $T_0 : B \to B$ be a mapping of class $L\,(k)$ with $d\,(T_0) = d > 0$. Then, for any $x \in B$ with $\|x\| \geq \sqrt{1 - d^2}$, we have*

$$0 \leq \|x\|^2 + d^2 - 1 \leq \|x\|^2 + \|T_0 x - x\|^2 - 1$$
$$= (x, x) + \|T_0 x\|^2 - (x, T_0 x) - (T_0 x - x, x) - 1$$
$$= -2\,(T_0 x - x, x) + \|T_0 x\|^2 - 1 \leq -2\,(T_0 x - x, x).$$

Consequently,

$$(T_0 x - x, x) \leq 0. \tag{9.4}$$

Let us modify T_0 defining $T_1 : B \to B$ as

$$T_1 x = \begin{cases} T_0 x & \text{if } \|x\| \leq \sqrt{1 - d^2}, \\ T_0 P_{\sqrt{1 - d^2}} x & \text{if } \sqrt{1 - d^2} \leq \|x\| \leq 1, \end{cases}$$

where $P_{\sqrt{1 - d^2}}$ is the radial projection onto the ball $\sqrt{1 - d^2}B$. Since $P_{\sqrt{1 - d^2}}$ is nonexpansive, T_1 is also of class $L\,(k)$. We shall prove that, for any $x \in S$ and for any $\mu \in [0, 1]$ and $\lambda \geq 1$,

$$\|\mu T_1 x - \lambda x\| \geq d. \tag{9.5}$$

Indeed, using (9.4), we get

$$\|\mu T_0 P x - \lambda x\|^2 = \left\| \mu T_0 \left(\sqrt{1 - d^2} x \right) - \lambda x \right\|^2$$

$$= \left\| \mu T_0 \left(\sqrt{1 - d^2} x \right) - \mu \sqrt{1 - d^2} x + \left(\mu \sqrt{1 - d^2} - \lambda \right) x \right\|^2$$

$$= \mu^2 \left\| T_0 \left(\sqrt{1 - d^2} x \right) - \sqrt{1 - d^2} x \right\|^2$$

$$+ 2 \frac{\mu \left(\mu \sqrt{1 - d^2} - \lambda \right)}{\sqrt{1 - d^2}} \left(T_0 \left(\sqrt{1 - d^2} x \right) - \sqrt{1 - d^2} x, \sqrt{1 - d^2} x \right)$$

$$+ \left(\mu \sqrt{1 - d^2} - \lambda \right)^2 \|x\|^2$$

$$\geq \mu^2 d^2 + \left(\mu\sqrt{1-d^2} - \lambda\right)^2$$

$$\geq d^2.$$

The above calculations also imply that, for any $x \in B$ *with* $\|x\| \in \left[\sqrt{1-d^2}, 1\right]$, *we have* $\|T_1 x - x\| \geq d$ *and for* $x \in S$,

$$\|x - T_1 x\| \geq \sqrt{2\left(1 - \sqrt{1-d^2}\right)} > d.$$

In particular, the above implies that $d(T_1) \geq d(T_0) = d > 0$.
 Now, let us define the domain

$$D = \left\{ x \in H : \|x\| \leq 1 + \frac{1}{k} - \frac{\|T_1 P x\|}{k} \right\}$$

and extend T_1 *to the mapping* $T_2 : D \to B$ *putting*

$$T_2 x = \begin{cases} T_1 x & \text{for } x \in B, \\ T_1 P x & \text{for } x \in D \setminus B, \end{cases}$$

where P *denotes the radial projection of* H *onto* B. *Obviously,* $T_2 \in L(k)$ *and, in view of (9.5),* $d(T_2) \geq d$. *Again, let us extend* T_2 *defining the mapping* $T_3 : D \cup \left(1 + \frac{1}{k}\right) S \to B$ *as*

$$T_3 x = \begin{cases} T_2 x & \text{if } x \in D, \\ 0 & \text{if } x \in \left(1 + \frac{1}{k}\right) S. \end{cases}$$

It is clear that T_3 *is still of class* $L(k)$ *and has* $d(T_3) = d(T_2) \geq d > 0$. *In view of the Kirzbraun-Valentine Theorem,* T_3 *can be once more extended to the mapping* $T_4 : \left(1 + \frac{1}{k}\right) B \to B$ *of class* $L(k)$. *Observe that for every* $x \in S$ *such an extension must map the segment joining points* $\left(1 + \frac{1}{k} - \frac{\|T_1 P x\|}{k}\right) x$ *and* $\left(1 + \frac{1}{k}\right) x$, *of length* $\frac{\|T_1 P x\|}{k}$, *onto an arc joining* $T_1 P x$ *and* 0 *of length not less than* $\|T_1 P x\|$. *Thus, this arc must be the segment. Consequently, an extension can be done in exactly one way:*

$$T_4 x = \begin{cases} T_2 x & \text{if } x \in D, \\ k\left(1 + \frac{1}{k} - \|x\|\right) \frac{T_1 P x}{\|T_1 P x\|} & \text{if } x \in \left(1 + \frac{1}{k}\right) B \setminus D. \end{cases}$$

According to (9.5), we have $d(T_4) \geq d$. *Now, we can define* $T : B \to B$ *of class* $L(k)$ *with* $T(S) = \{0\}$ *as*

$$T x = \frac{k}{k+1} T_4 \left(\frac{k+1}{k} x\right).$$

Obviously, $d(T) \geq \frac{k}{k+1} d$, *and, consequently, since* d *can be chosen arbitrarily close to* $\psi_H(k)$, *we get*

$$\psi_{S \to \{0\}}(k) \geq \frac{k}{k+1} \psi_H(k). \tag{9.6}$$

It is not known whether the above estimate is accurate.

9.4 Optimal retractions

The functions ψ_X, $\psi_{B \to S}$ and $\psi_{S \to \{0\}}$ are closely related to another highly nontrivial problem posed by Goebel in 1973, which, roughly speaking, consists in constructing a retraction of the closed unit ball onto the unit sphere with the smallest possible Lipschitz constant. More precisely, for a given infinitely dimensional Banach space X, we define the *optimal retraction constant*:

$$k_0 (X) = \inf \{k : \text{ there exists a retraction } R : B \to S \text{ of class } L(k)\}.$$

The optimal retraction problem involves finding or evaluating the constant $k_0 (X)$. At present, the exact value of $k_0(X)$ is not known for any single Banach space. It is a general feeling that a value of the constant $k_0 (X)$ should somehow depend on the geometry of the space X, and its value should increase as the geometry gets a more "regular" structure.

There were several approaches to give a reasonable universal estimate from above. All of them ended on the level of the high thousands. Much better estimates exist for particular spaces or some classes of spaces. All of them have been obtained by a range of individual methods, constructions, and tricks developed by a number of authors. Below, we present the best of currently known estimates.

Let us begin with some basic estimates from below.

Suppose $R : B \to S$ is a retraction of class $L(k)$. Consider the mapping $T = -R$ and observe that $T : B \to S$, $T^n = (-1)^n R$, and this implies that T is uniformly lipschitzian, $T \in L_u(k)$. Moreover, T is fixed point free. Take any point $x \in B$ and consider the segment $[x, Tx]$. The mapping T maps this segment onto a rectifiable curve γ contained in S and joining two antipodal points $Tx = -Rx$ and $T^2x = Rx$ via relation

$$\gamma(s) = T\left((1-s)x + sTx\right),$$

where $s \in [0, 1]$. The length of γ, defined as

$$l(\gamma) = \sup \sum_{i=0}^{n-1} \|\gamma(s_i) - \gamma(s_{i+1})\|,$$

where the supremum is taken over all finite partitions $0 = s_0 < s_1 < s_2 < \cdots < s_n = 1$ of $[0, 1]$, exceeds *the girth* $g(X)$ of the sphere defined as the infimum of lengths of all curves joining two antipodal points.

On the other hand, since $T \in L(k)$ we get

$$l(\gamma) \le \sup \sum_{i=0}^{n-1} \|T\left((1-s_i)x + s_iTx\right) - T\left((1-s_{i+1})x + s_{i+1}Tx\right)\|$$

$$\leq \sup \sum_{i=0}^{n-1} k\left(s_{i+1} - s_i\right)\|x - Tx\| = k\|x - Tx\|.$$

The above implies that $l(\gamma)$ satisfies

$$g(X) \leq l(\gamma) \leq k\|x - Tx\|,$$

and, consequently, the minimal displacement of T is positive,

$$d(T) \geq \frac{g(X)}{k} > 0.$$

Further, since $d(T) \leq \psi_{B \to S}(k)$, we get the inequality, which can be used to obtain some basic estimates for $k_0(X)$:

$$k\psi_{B \to S}(k) \geq g(X). \tag{9.7}$$

Obviously, $g(X) \geq 2$ for all spaces X, and there are some flat spaces for which $g(X) = 2$. Since $\psi_{B \to S}(k) \leq \psi(k) \leq 1 - \frac{1}{k}$ we get the first basic estimate for the optimal retraction constant, $k_0(X) \geq 3$ for any space X. However, for some spaces there are better estimates. For the space ℓ_1, we have $g(\ell_1) = 2$ and by using the estimate (9.1) to solve the inequality (9.7), we get

$$k_0(\ell_1) \geq 17 - 8\sqrt{3} = 3.143\cdots.$$

A better result has been obtained in [12], and it states that $k_0(\ell_1) \geq 4$. It is known that for all uniformly convex spaces $g(X) > 2$, and, consequently, $k_0(X) > 3$ for every such space. In the particular case of Hilbert space H, $g(H) = \pi$. Having used this fact and the evaluation (9.2) for $\psi_{B \to S}$ presented in example 9.12, we conclude that $k_0(H)$ exceeds the solution of the equation

$$k\left(1 - \frac{1}{k}\right)^{\frac{3}{2}} = \pi.$$

Thus, $k_0(H) > 4.5\cdots$. For details concerning the girth of the sphere, see [73].

Considerably more published results concern estimates from above. All the results of this type have been obtained via concrete examples of mappings.

Let us begin with a construction in the space ℓ_1 presented by Annoni and Casini in [2].

Example 9.16 *Consider the space ℓ_1 furnished with the standard norm. Let $i : \overline{B \setminus \frac{1}{2}B} \to \mathbb{N}$ be the function defined for $x = (x_1, x_2, \dots) \in \ell_1$ with $\|x\| \in \left[\frac{1}{2}, 1\right]$ as*

$$i(x) = \min\left\{ j \in \mathbb{N} : \sum_{k=j+1}^{\infty} |x_k| < 1 - \|x\| \right\}$$

and let the map $\mu : \overline{B \setminus \frac{1}{2}B} \to (0,1]$ *be given via relation*

$$\mu(x)\left|x_{i(x)}\right| + \sum_{k=i(x)+1}^{\infty} |x_k| = 1 - \|x\|.$$

Now, if $\{e_k\}_{k=1}^{\infty}$ *denotes the standard Schauder basis in* ℓ_1, *we define* $Q : \overline{B \setminus \frac{1}{2}B} \to \frac{1}{2}B$ *by putting for every* $x = \sum_{k=1}^{\infty} x_k e_k$

$$Qx = \begin{cases} \mu(x)\,x_{i(x)}e_{i(x)} + \sum_{k=i(x)+1}^{\infty} x_k e_k & \text{if } \ \frac{1}{2} \le \|x\| < 1, \\ 0 & \text{if } \ \|x\| = 1. \end{cases}$$

Justification that Q *is 3-lipschitzian and* $I - Q$ *is 2-lipschitzian, where* I *denotes the identity map, is left for the reader as an exercise.*

We shall also need the "right shift" mapping $A : B \to B$,

$$Ax = A\left(\sum_{k=1}^{\infty} x_k e_k\right) = \sum_{k=1}^{\infty} x_k e_{k+1}.$$

Obviously, A *as an isometry is of class* $L(1)$.

Finally, we can define $R : B \to S$ *as*

$$Rx = \begin{cases} (1 - 2\|x\|)\,e_1 + 2Ax & \text{if } \ 0 \le \|x\| \le \frac{1}{2}, \\ (I - Q)x + 2AQx & \text{if } \ \frac{1}{2} \le \|x\| \le 1. \end{cases}$$

The reader can easily verify that R *is a retraction of* B *onto* S *and that* $R \in L(8)$. *Consequently,* $k_0(\ell_1) \le 8$, *and it is the best known upper bound for this space.*

Next, with the use of the same tricks, the above result was extended for $L_1(0,1)$ and for few other spaces (see [39]). Below, we present yet another simple construction in the space $L_1(0,1)$ given by the present author (unpublished), which is also based on an idea originated by Annoni and Casini.

Example 9.17 *Consider the space* $L_1(0,1)$. *For every function* $f \in B$, *we define the number* t_f *as*

$$t_f = \inf\left\{t \in (0,1) : \int_t^1 |f(t)|\,dt \le 1 - \|f\|\right\}.$$

If $\|f\| \le \frac{1}{2}$, *then* $t_f = 0$. *If* $\|f\| \in [\frac{1}{2}, 1]$, *then*

$$\int_{t_f}^1 |f(t)|\,dt = 1 - \|f\|.$$

Let us define a mapping $T : \overline{B \setminus \frac{1}{2}B} \to S$ *as*

$$(Tf)(t) = \begin{cases} f(t) & \text{if } t \in (0, t_f], \\ 2\,|f(t)| & \text{if } t \in (t_f, 1). \end{cases}$$

We shall prove that T is 8-lipschitzian. Let $f, g \in \overline{B \setminus \frac{1}{2}B}$, and without loss of generality, we can assume that $t_f \le t_g$. Then,

$$\|Tf - Tg\| = \int_0^{t_f} |f(t) - g(t)|\,dt + \int_{t_f}^{t_g} |2\,|f(t)| - g(t)|\,dt$$

$$+ \int_{t_g}^{1} |2\,|f(t)| - 2\,|g(t)||\,dt$$

$$\le \int_0^{t_f} |f(t) - g(t)|\,dt + \int_{t_f}^{t_g} |2\,|f(t)| - 2\,|g(t)||\,dt$$

$$+ 3\int_{t_f}^{t_g} |g(t)|\,dt + \int_{t_g}^{1} |2\,|f(t)| - 2\,|g(t)||\,dt$$

$$\le 2\,\|f - g\| + 3\left(\int_{t_f}^{1} |g(t)|\,dt - \int_{t_g}^{1} |g(t)|\,dt \right)$$

$$= 2\,\|f - g\| + 3\left(\int_{t_f}^{1} |g(t)|\,dt - 1 + \|g\| \right)$$

$$= 2\,\|f - g\| + 3\left(\int_{t_f}^{1} |g(t)|\,dt - \|f\| - \int_{t_f}^{1} |f(t)|\,dt + \|g\| \right)$$

$$\le 2\,\|f - g\| + 3\left(\int_{t_f}^{1} |f(t) - g(t)|\,dt + \|f - g\| \right)$$

$$\le 8\,\|f - g\|.$$

Now we can define the mapping $R : B \to S$ as

$$(Rf)(t) = \begin{cases} 2\,|f(t)| - 2\,\|f\| + 1 & \text{if } \|f\| \le \frac{1}{2}, \\ (Tf)(t) & \text{if } \|f\| \in [\frac{1}{2}, 1]. \end{cases}$$

and the reader can easily verify that R is an 8-lipschitzian retraction of the solid unit ball B onto its boundary S. Hence, $k_0\,(L_1\,(0, 1)) \le 8$, and it is the best known estimate for this space.

Let us pass now to the spaces furnished with the sup norm. For many years, the best known estimate for the space $C\,[0, 1]$ was $k_0\,(C\,[0, 1]) \le 4\,(1 + \sqrt{2})^2 = 23.31\cdots$ (see [31]). This estimate was obtained by the following procedure, which works in case of any Banach space X. Once we have the mapping $T : B_X \to B_X$ of class $T \in L\,(k)$ such that $d\,(T) > 0$ and $T\,(S_X) = \{0\}$, we can construct the retraction $R : B_X \to S_X$ by putting

$$Rx = \frac{x - Tx}{\|x - Tx\|} = P\left(\frac{x - Tx}{d(T)}\right),$$

where P denotes the radial projection. Then,

$$k(R) \le k(P)\frac{k+1}{d(T)}.$$

Since the mapping T can be chosen having $d(T)$ arbitrarily close to $\psi_{S_X \to \{0\}}(k)$, we conclude that

$$k_0(X) \le k(P) \inf_{k>1} \frac{k+1}{\psi_{S_X \to \{0\}}(k)}. \tag{9.8}$$

We have already seen in example 9.13 that for $X = C[0,1]$, and in general, for all extremal and truncation invariant subspaces X of $B(K)$,

$$\psi_{S_X \to \{0\}}(k) \ge \frac{1}{2}\left(1 - \frac{1}{k}\right) = \frac{1}{2}\psi_X(k).$$

Thus, for any such space (see [39])

$$3 \le k_0(X) \le 4 \min_{k>1} \frac{(k+1)k}{k-1} = 4\left(1 + \sqrt{2}\right)^2 = 23.31\cdots.$$

As we have already mentioned in the previous section, it was proved in [38] that for the space $C_0[0,1]$ we have $\psi_{S \to \{0\}}(3) = \frac{2}{3}$ and, consequently, $k_0(C_0[0,1]) \le 12$. Later, a better estimate was obtained in [40], $k_0(C_0[0,1]) \le 7$.

A disadvantage of the above method is the fact that for spaces c, c_0, $BC(\mathbb{R})$, $BC_0(\mathbb{R})$, $C[0,1]$, $C_0[0,1]$ and many others we have $k(P) = 2$; hence, the biggest possible Lipschitz constant for the radial projection.

However, it was observed by the present author in [67] that, in the case of the above-mentioned spaces, we can omit the radial projection P by applying the following receipt.

Again, suppose that X is an infinite-dimensional and truncation invariant subspace of $B(K)$. Consider a mapping $T : B_X \to B_X$ of class $L(k)$ such that $d(T) = d > 0$ and $T(S_X) = \{0\}$. Denote

$$r = \frac{k + 1 + \sqrt{(k+1)^2 - 4kd}}{2k}$$

and define the mapping $F : rB_X \to X$ as

$$Ff = \begin{cases} f - Tf & \text{if} \quad \|f\| \le 1, \\ (k + 1 - k\|f\|)f & \text{if} \quad \|f\| \in [1, r]. \end{cases}$$

It is easy to verify that, for any $f \in rB_X$, we have $\|Ff\| \geq d$. Moreover, the sphere with a radius r is mapped radially onto the sphere with a radius d; that is, for every $f \in rS_X$, we have $Ff = \frac{d}{r}f$. For a convenience of the reader we shall prove that $F \in L(k+1)$. It is clear that F is $(k+1)$-lipschitzian on B_X. To see that it also holds on the subset $rB_X \setminus B_X$, first assume that $\|g\| \leq \|f\|$. Then,

$$
\begin{aligned}
\|Ff - Fg\| &= \|(k+1 - k\|f\|) f - (k+1 - k\|g\|) g\| \\
&\leq \|(k+1 - k\|f\|)(f - g)\| \\
&\quad + \|(k+1 - k\|f\|) g - (k+1 - k\|g\|) g\| \\
&\leq (k+1 - k\|f\|) \|f - g\| + k\|g\| (\|f\| - \|g\|) \\
&\leq (k+1 - k\|f\| + k\|g\|) \|f - g\| \\
&\leq (k+1) \|f - g\|.
\end{aligned}
$$

If $\|f\| \leq \|g\|$, then it is enough to exchange roles between f and g in the above estimate. Finally, for all $f, g \in rB_X$, we have

$$
\|Ff - Fg\| \leq (k+1) \|f - g\|.
$$

Now define the mapping $\overline{F} : B_X \to X$ as

$$
\overline{F}f = \frac{1}{r} F(rf).
$$

We leave to the reader checking that

- $\overline{F} \in L(k+1)$;

- for any $f \in B_X$ we have $\|\overline{F}f\| \geq \frac{d}{r}$;

- for every $f \in S_X$ we have $\overline{F}f = \frac{d}{r}f$.

Finally, we can define the retraction $R : B_X \to S_X$ by putting

$$
Rf = Q^\tau \left(\frac{r}{d} \overline{F}f \right)
$$

and observe that

$$
\begin{aligned}
\|Rf - Rg\| &= \left\| Q^\tau \left(\frac{r}{d} \overline{F}f \right) - Q^\tau \left(\frac{r}{d} \overline{F}g \right) \right\| \leq \left\| \frac{r}{d} \overline{F}f - \frac{r}{d} \overline{F}g \right\| \\
&\leq \frac{r}{d} (k+1) \|f - g\|.
\end{aligned}
$$

Consequently, since T can be selected with $d(T)$ arbitrarily close to $\psi_{S_X \to \{0\}}(k)$, we obtain the following

Theorem 9.14 ([67]) *If X is an infinite-dimensional and truncation invariant subspace of $B(K)$, then*

$$
k_0(X) \leq \inf_{k>1} (k+1) \frac{k + 1 + \sqrt{(k+1)^2 - 4k\psi_{S_X \to \{0\}}(k)}}{2k\psi_{S_X \to \{0\}}(k)}.
$$

Example 9.18 *(see [69] or [67]) Consider an extremal and truncation invariant subspace X of $B(K)$. We have already seen in example 9.13 that*

$$\psi_{S_X \to \{0\}}(k) \geq \frac{1}{2}\left(1 - \frac{1}{k}\right).$$

Thus, in virtue of theorem 9.14, we get the estimate

$$k_0(X) \leq \min_{k>1}(k+1)\frac{k+1+\sqrt{(k+1)^2 - 2k(1-\frac{1}{k})}}{k(1-\frac{1}{k})}$$

$$= \min_{k>1}(k+1)\frac{k+1+\sqrt{k^2+3}}{k-1}.$$

We leave to the reader checking that the minimum is achieved for $k = 3$. Consequently,

$$k_0(X) \leq 4(2+\sqrt{3}) = 14.92\cdots.$$

We have already seen that among such spaces there are, for instance, c, c_0, $C[0,1]$, and $BC(\mathbb{R})$.

Example 9.19 *(see [67] or [68]) Consider the space $C_0[0,1]$. A construction presented in example 9.14 shows that*

$$\psi_{S \to \{0\}}(k) \geq \frac{k+1}{4}\left(1 - \frac{1}{k}\right) \quad \text{for } k \in [1,3]$$

and

$$\psi_{S \to \{0\}}(k) = 1 - \frac{1}{k} \quad \text{for } k \geq 3.$$

Consequently, the expression in theorem 9.14 takes its minimal value for $k = 3$. Thus, we have

$$3 \leq k_0(C_0[0,1]) \leq 2(2+\sqrt{2}) = 6.828\cdots.$$

At present the above estimate is the minimum of upper bounds over all the Banach spaces for which the upper bound is known!

The above construction can be extended to a much wider class of spaces.

Example 9.20 *(see [67] or [68]) Suppose that (M,d) is a connected metric space consisting of more than one point and $z \in M$ is a given point. Consider the space $BC_z(M)$ consisting of all bounded, continuous functions $f : M \to \mathbb{R}$ vanishing at z, $f(z) = 0$, and furnished with the standard sup norm.*
First, assume that (M,d) is unbounded and define T as

$$Tf(x) = \Lambda\left(3\left(|f(x)| + \frac{d(x,z)}{1+d(x,z)}\right)\right).$$

Then, $T : B \to B$ and $T \in L(3)$. Since M is a connected metric space, for every $f \in B$ there exists a point $x_1 \in M$ such that

$$|f(x_1)| + \frac{d(x_1, z)}{1 + d(x_1, z)} = \frac{1}{3},$$

and, consequently, $\|f - Tf\| \geq \frac{2}{3}$.

In the next step, define a mapping $F : \left(\frac{2+\sqrt{2}}{3}\right) B \to BC_z(M)$ as

$$Ff = \begin{cases} f - Tf & \text{if } \|f\| \leq \frac{2}{3}, \\ f - Q^\tau_{3(1-\|f\|)} Tf & \text{if } \|f\| \in \left[\frac{2}{3}, 1\right], \\ (4 - 3\|f\|) f & \text{if } \|f\| \in \left[1, \frac{2+\sqrt{2}}{3}\right]. \end{cases}$$

We leave to the reader to check that $F \in L(4)$ and $\|Ff\| \geq \frac{2}{3}$ for any $f \in \left(\frac{2+\sqrt{2}}{3}\right) B$.

Now, having denoted

$$\tilde{F}f = \frac{3}{2+\sqrt{2}} F\left(\frac{2+\sqrt{2}}{3} f\right),$$

we obtain a mapping $\tilde{F} : B \to BC_z(M)$ of class $L(4)$ such that, for every $f \in B$, we have $\left\|\tilde{F}f\right\| \geq \frac{2}{2+\sqrt{2}}$ and for every $f \in S$, $\tilde{F}f = \frac{2}{2+\sqrt{2}} f$.

Finally, we can define the retraction $R : B \to S$ as

$$Rf = Q^\tau \left(\frac{2+\sqrt{2}}{2} \tilde{F}f\right)$$

and observe that for all $f, g \in B$ we have

$$\|Rf - Rg\| \leq \frac{2+\sqrt{2}}{2} \left\|\tilde{F}f - \tilde{F}g\right\| \leq 2\left(2 + \sqrt{2}\right) \|f - g\|.$$

If (M, d) is bounded, then it is enough to put

$$m = \sup\{d(x, z) : x \in M\}$$

and modify T by putting

$$Tf(x) = \Lambda\left(3\left(|f(x)| + \frac{d(x, z)}{m}\right)\right).$$

The proof carries on with only minor technical changes. All the above implies that

$$3 \leq k_0\left(BC_z(M)\right) \leq 2\left(2 + \sqrt{2}\right) = 6.828\cdots.$$

It is worth noting that for the whole class of extremal spaces the best known estimate from above states $k_0(X) \leq 30.84$ (see [6]).

The case of Hilbert space H is very challenging. Because of its geometric regularity, it opposes the examination of minimal displacement and optimal retraction.

The first upper bound $k_0(H) \leq 64.25$ was obtained by Komorowski and Wośko in [53]. Subsequently, there were several improvements, for instance, $k_0(H) \leq 32.26$ in [13] and $k_0(H) \leq 28.99$ from [6]. The last estimate is the best known published estimate for the Hilbert space.

There are some relations that tie the unknown value of $k_0(H)$ and the unknown function ψ_H. Since, in this case $k(P) = 1$, then in view of (9.8) and (9.6), we have

$$k_0(H) \leq \inf_{k>1} \frac{k+1}{\psi_{S \to \{0\}}(k)} \leq \inf_{k>1} \frac{(k+1)^2}{k\psi_H(k)}.$$

Since for any space $\psi(k) \geq \psi'(1)\left(1 - \frac{1}{k}\right)$, we get

$$k_0(H)\,\psi'_H(1) \leq \inf_{k>1} \frac{(k+1)^2}{k-1} = 8.$$

It is not known if the above estimates are sharp.

In spite of a forty-year efforts, the quantitative problems of minimal displacement and optimal retraction are still open for consideration. We hold a deep-seated belief that the area needs new ideas and approaches.

9.5   Generalized characteristics of minimal displacement

By analogy to the classical case, we define minimal displacement functions for classes $L(\alpha, p, k)$. As usual, C denotes a bounded, closed, convex, and noncompact subset of an infinitely dimensional Banach space X with $r(C) = 1$. By φ_C, we denote a function defined for every $k \geq 0$ and $\alpha = (\alpha_1, \dots, \alpha_n)$ by

$$\varphi_C(\alpha, k) = \sup\{d(T) : T : C \to C, \; T \in L(\alpha, k)\}$$
$$= \sup\{\inf\{\|x - Tx\| : x \in C\} : T : C \to C, \; T \in L(\alpha, k)\}$$

and by φ_X a function given for every $k \geq 0$ and $\alpha = (\alpha_1, \dots, \alpha_n)$ by

$$\varphi_X(\alpha, k) = \sup\{\varphi_C(\alpha, k) : C \subset X, \; r(C) = 1\}.$$

If $C = B_X$, then we write ψ_X instead of φ_{B_X}, and we refer to ψ_X as *the minimal displacement characteristic of X for mappings in the class $L(\alpha, k)$.*

Thus,
$$\psi_X(\alpha, k) = \sup\{d(T): \ T: B_X \to B_X, \ T \in L(\alpha, k)\}.$$
Similarly, for classes $L(\alpha, p, k)$, we define functions φ_C, φ_X and ψ_X as
$$\varphi_C(\alpha, p, k) = \sup\{d(T): \ T: C \to C, \ T \in L(\alpha, p, k)\},$$
$$\varphi_X(\alpha, p, k) = \sup\{\varphi_C(\alpha, p, k): \ C \subset X, \ r(C) = 1\}$$
and
$$\psi_X(\alpha, p, k) = \sup\{d(T): \ T: B_X \to B_X, \ T \in L(\alpha, p, k)\}.$$

If $p = 1$, then we identify $\varphi_C(\alpha, 1, k)$ with $\varphi_C(\alpha, k)$. The same holds for $\psi_X(\alpha, 1, k)$ and $\varphi_X(\alpha, 1, k)$. It is easy to see that for a multi-index $\alpha = (1)$ of length $n = 1$ the above definitions coincide with the classical definitions of characteristics of minimal displacement; that is,
$$\varphi_C(\alpha, p, k) = \varphi_C(k) = \sup\{d(T): \ T: C \to C, \ T \in L(k)\},$$
$$\varphi_X(\alpha, p, k) = \varphi_X(k) = \sup\{\varphi_C(k): \ C \subset X, \ r(C) = 1\}$$
and
$$\psi_X(\alpha, p, k) = \psi_X(k) = \sup\{d(T): \ T: B_X \to B_X, \ T \in L(k)\}.$$

Let us list some immediate consequences of the above definitions:

- Since $L(\alpha, p, k) \subset L(\alpha, p, l)$ for $k < l$, we conclude that for fixed α and p, functions φ_C, φ_X and ψ_X are nondecreasing.

- Since $L(\alpha, p, k) \subset L(\alpha, q, k)$ for $p > q$, we conclude that for fixed α and k functions φ_C, φ_X and ψ_X are nonincreasing.

- For any α, p, and k, we have
$$\varphi_C(\alpha, p, k) \le \varphi_X(\alpha, p, k).$$
In particular, $\psi_X(\alpha, p, k) \le \varphi_X(\alpha, p, k)$.

- In view of theorem 3.9, for any α, p, and $k < 1$, we have $\varphi_C(\alpha, p, k) = 0$. Consequently, for any α, p, and $k < 1$, we have
$$\varphi_X(\alpha, p, k) = \psi_X(\alpha, p, k) = 0.$$

- Using (5.2), for any $\alpha = (\alpha_1, \ldots, \alpha_n)$, p and k, the class $L(\alpha, p, k)$ contains the class $L(l)$ with l satisfying the following equation:
$$\sum_{i=1}^{n} \alpha_i l^{ip} = k^p.$$
It is clear that $l > 1$ provided $k > 1$. Thus, in view of theorem 9.11, for any $k > 1$, we have
$$\varphi_C(\alpha, p, k) \ge \varphi_C(l) > 0.$$

The first evaluation of the minimal displacement characteristic of c_0 for mappings in the class $L(\alpha, k)$ was presented in [42]:

Example 9.21 *Let c_0 denotes the space of all sequences converging to zero with the standard sup norm, $\|x\| = \|(x_1, x_2, \dots)\| = \sup_{i=1,2,\dots} |x_i|$. Let $\tau : \mathbb{R} \to [-1, 1]$ be the truncation function as in example 9.4.*

For arbitrary $k > 1$, let us define the mapping $T : B_{c_0} \to B_{c_0}$ by

$$Tx = T(x_1, x_2, \dots) = (1, \tau(kx_1), \tau(kx_2), \dots).$$

We leave to the reader the verification that T is lipschitzian with $k(T) = k$ and for all $x \in B_{c_0}$ we have $\|x - Tx\| > 1 - 1/k$.

The space c_0 is isometric to the product $c_0 \times c_0$ with the maximum norm. The unit ball in this setting is the product of two unit balls B_{c_0}. Let us define the mapping $F : B_{c_0} \times B_{c_0} \to B_{c_0} \times B_{c_0}$ by

$$F(x, y) = (y, Tx).$$

Then,

$$F^2(x, y) = (Tx, Ty),\ F^3(x, y) = (Ty, T^2x),\ F^4(x, y) = (T^2x, T^2y),$$
$$F^5(x, y) = (T^2y, T^3x),\ F^6(x, y) = (T^3x, T^3y), \dots.$$

Thus,

$$k(F) = k(F^2) = k,\ k(F^3) = k(F^4) = k^2,\ k(F^5) = k(F^6) = k^3, \dots.$$

Consequently, for any multi-index α of length 2, $F \in L(\alpha, k)$. Further, for all $(x, y) \in B_{c_0} \times B_{c_0}$, we have

$$
\begin{aligned}
\|F(x, y) - (x, y)\| &= \|(y, Tx) - (x, y)\| \\
&= \max\{\|y - x\|, \|Tx - y\|\} \\
&\geq \max\{\|y - x\|, \|Tx - x\| - \|y - x\|\} \\
&\geq \max\{\|y - x\|, 1 - 1/k - \|y - x\|\} \\
&\geq 1/2\,(1 - 1/k).
\end{aligned}
$$

Thus, for all α of length 2,

$$\psi_{c_0}(\alpha, k) \geq \frac{1}{2}\psi_{c_0}(k) = \frac{1}{2}\left(1 - \frac{1}{k}\right).$$

Fix a multi-index $\alpha = (\alpha_1, \dots, \alpha_n)$ and $k > 1$. Let $w : \mathbb{R} \to \mathbb{R}$ be defined by

$$w(t) = \alpha_1 t + \alpha_2 t^2 + \cdots + \alpha_n t^n$$

and l (more precisely $l = l(\alpha, k)$) denotes the unique solution of the equation

$$w(t) = k \text{ for } t > 1.$$

Hence, $l > 1$ and

$$w(l) = \alpha_1 l + \alpha_2 l^2 + \cdots + \alpha_n l^n = k.$$

Under the above notations, we formulate the following:

Theorem 9.15 *For every* $\alpha = (\alpha_1, \ldots, \alpha_n)$ *and* $k > 1$, *we have*

$$\varphi'_C(1) \left(1 - \frac{1}{l} \right) \leq \varphi_C(\alpha, k) \leq \varphi_C(k/\alpha_1) \leq 1 - \frac{\alpha_1}{k}, \tag{9.9}$$

$$\psi'_X(1) \left(1 - \frac{1}{l} \right) \leq \psi_X(\alpha, k) \leq \psi_X(k/\alpha_1) \leq 1 - \frac{\alpha_1}{k}$$

and

$$\varphi'_X(1) \left(1 - \frac{1}{l} \right) \leq \varphi_X(\alpha, k) \leq \varphi_X(k/\alpha_1) \leq 1 - \frac{\alpha_1}{k}.$$

Proof. We shall prove inequality (9.9). Using the fact that $L(\alpha, k) \subset L(k/\alpha_1)$ and lemma 9.4, we get the right-hand side of (9.9). To prove the left-hand side of (9.9), we observe that, in view of (3.10), $L(l) \subset L(\alpha, k)$. Thus, $\varphi_C(\alpha, k) \geq \varphi_C(l)$, and from lemma 9.4, we finally obtain

$$\varphi_C(\alpha, k) \geq \varphi_C(l) \geq \varphi'_C(1) \left(1 - \frac{1}{l} \right).$$

$\square$

A disadvantage of theorem 9.15 is that we have to determine the root l of polynomial $w(t) - k$. We can remedy this problem by making the above inequality weaker but much easier to determine, as seen below:

Theorem 9.16 *For any* $\alpha = (\alpha_1, \ldots, \alpha_n)$ *and* $k > 1$, *we have*

$$\varphi_C(\alpha, k) \geq \frac{\varphi'_C(1)}{w'(k)} \left(1 - \frac{1}{k} \right),$$

$$\psi_X(\alpha, k) \geq \frac{\psi'_X(1)}{w'(k)} \left(1 - \frac{1}{k} \right)$$

and

$$\varphi_X(\alpha, k) \geq \frac{\varphi'_X(1)}{w'(k)} \left(1 - \frac{1}{k} \right).$$

Proof. Similarly to the previous case, we shall prove only the first inequality. Using (3.10), we get $L(l) \subset L(\alpha, k)$. Thus, in view of lemma 9.4, for $s < l$, we can write

$$\varphi_C(\alpha, k) \geq \varphi_C(l) \geq \frac{s \varphi_C(s)}{s - 1} \cdot \frac{l - 1}{w(l) - 1} \cdot \frac{k}{l} \cdot \frac{w(l) - 1}{k}.$$

Using lemma 9.4 and the fact that $k \geq l$, we obtain

$$\varphi_C(\alpha, k) \geq \varphi'_C(1) \cdot \frac{l-1}{w(l)-1} \cdot \left(1 - \frac{1}{k}\right).$$

In view of Lagrange's Mean Value Theorem, we have

$$\frac{w(l)-1}{l-1} = w'(\theta) \text{ for some } \theta \in (1, l).$$

Since $w'(\theta) \leq w'(l) \leq w'(k)$, we finally get

$$\varphi_C(\alpha, k) \geq \frac{\varphi'_C(1)}{w'(k)} \left(1 - \frac{1}{k}\right).$$

$\square$

In case of multi-index α of length $n = 2$, we can give better evaluations:

Theorem 9.17 *For any $\alpha = (\alpha_1, \alpha_2)$ and $k > 1$, we have*

$$\varphi_C(\alpha, k) \geq \frac{\varphi'_C(1)}{w'(1)} \left(1 - \frac{1}{k}\right), \tag{9.10}$$

$$\psi_X(\alpha, k) \geq \frac{\psi'_X(1)}{w'(1)} \left(1 - \frac{1}{k}\right),$$

and

$$\varphi_X(\alpha, k) \geq \frac{\varphi'_X(1)}{w'(1)} \left(1 - \frac{1}{k}\right).$$

Proof. We start as in the proof of the previous theorem,

$$\varphi_C(\alpha, k) \geq \varphi'_C(1) \cdot \frac{l-1}{w(l)-1} \cdot \frac{w(l)}{l} \cdot \frac{w(l)-1}{w(l)}.$$

To get the thesis, it is enough to show that, for any $s > 1$, we have

$$\frac{s-1}{w(s)-1} \cdot \frac{w(s)}{s} \geq \frac{1}{w'(1)}, \tag{9.11}$$

or equivalently,

$$(w'(1) - 1) w(s) - w'(1)\frac{w(s)}{s} + 1 \geq 0.$$

Indeed, the left-hand side of the last inequality is equal to

$$(\alpha_1 + 2\alpha_2 - 1)(\alpha_1 s + \alpha_2 s^2) - (\alpha_1 + 2\alpha_2)(\alpha_1 + \alpha_2 s) + 1$$

and since $\alpha_1 + 2\alpha_2 - 1 = \alpha_2$ and $\alpha_1 + 2\alpha_2 = 1 + \alpha_2$, the above is equal to

$$\alpha_2 \left(\alpha_1 s + \alpha_2 s^2\right) - (1 + \alpha_2)(\alpha_1 + \alpha_2 s) + 1$$

$$= \alpha_1 \alpha_2 s + \alpha_2^2 s^2 - \alpha_1 - \alpha_1 \alpha_2 - \alpha_2 s - \alpha_2^2 s + 1$$

$$= \alpha_2^2 s^2 + \left(\alpha_1 \alpha_2 - \alpha_2 - \alpha_2^2\right) s + 1 - \alpha_1 - \alpha_1 \alpha_2$$

$$= \alpha_2^2 s^2 + \left[\alpha_2(\alpha_1 - 1) - \alpha_2^2\right] s + \alpha_2(1 - \alpha_1)$$

$$= \alpha_2^2 s^2 - 2\alpha_2^2 s + \alpha_2^2$$

$$= \alpha_2^2 (s - 1)^2 \geq 0$$

as desired.

$\square$

In general, the evaluation (9.11) does not hold; that is, it is not true that, for any multi-index $\alpha = (\alpha_1, \ldots, \alpha_n)$ and $s > 1$, we have

$$\frac{s-1}{w(s)-1} \cdot \frac{w(s)}{s} \geq \frac{1}{w'(1)}.$$

To see this, consider, for example,

$$\alpha = \left(\frac{98}{100}, \frac{1}{100}, \frac{1}{100}\right) \quad \text{or} \quad \alpha = \left(\frac{3}{4}, \frac{1}{12}, \frac{1}{12}, \frac{1}{12}\right).$$

Nevertheless, it is easy to verify that, for any multi-index $\alpha = (\alpha_1, \alpha_2)$ of length $n = 2$ and $k > 1$, we obtain

$$l = \frac{2k}{\alpha_1 + \sqrt{\alpha_1^2 + 4\alpha_2 k}}.$$

Hence, in view of lemma 9.4,

$$\varphi_C(\alpha, k) \geq \varphi_C'(1) \left(1 - \frac{\alpha_1 + \sqrt{\alpha_1^2 + 4\alpha_2 k}}{2k}\right),$$

which is better bound than (9.10). Moreover, in the case of extremal space X, we have $\psi_X'(1) = 1$, and, consequently,

$$1 - \frac{\left(\alpha_1 + \sqrt{\alpha_1^2 + 4\alpha_2 k}\right)}{2k} \leq \psi_X(\alpha, k) \leq 1 - \frac{\alpha_1}{k}. \tag{9.12}$$

The following theorem was proved in [42].

Theorem 9.18 *For any* $\alpha = (\alpha_1, \ldots, \alpha_n)$ *we have* $\displaystyle\lim_{k \to \infty} \psi_X(\alpha, k) = 1.$

Proof. It is clear that

$$\psi_X(l) \le \psi_X(\alpha, k) \le \psi_X(k/\alpha_1).$$

If $k \to \infty$ (for a fixed α), then $l \to \infty$. Hence, in view of lemma 9.4, we get

$$\lim_{k \to \infty} \psi_X(l) = \lim_{k \to \infty} \psi_X(k/\alpha_1) = 1.$$

Thus, $\lim_{k \to \infty} \psi_X(\alpha, k) = 1.$

$\square$

From the above theorem, we conclude that the evaluation presented in example 9.21 is not sharp.

The above results can be extended easily for classes $L(\alpha, p, k)$.

Theorem 9.19 *For any $\alpha = (\alpha_1, \dots, \alpha_n)$, $p \ge 1$, and $k > 1$, we have*

$$\varphi'_C(1) \left(1 - \frac{1}{l} \right) \le \varphi_C(\alpha, p, k) \le \varphi_C(k/\alpha_1^{1/p}) \le 1 - \frac{\alpha_1^{1/p}}{k}, \qquad (9.13)$$

$$\psi'_X(1) \left(1 - \frac{1}{l} \right) \le \psi_X(\alpha, p, k) \le \psi_X(k/\alpha_1^{1/p}) \le 1 - \frac{\alpha_1^{1/p}}{k},$$

and

$$\varphi'_X(1) \left(1 - \frac{1}{l} \right) \le \varphi_X(\alpha, p, k) \le \varphi_X(k/\alpha_1^{1/p}) \le 1 - \frac{\alpha_1^{1/p}}{k},$$

where $l = l(\alpha, p, k) > 1$ satisfies the condition

$$w(l^p) = \alpha_1 l^p + \alpha_2 l^{2p} + \cdots + \alpha_n l^{np} = k^p. \qquad (9.14)$$

Proof. Using the fact that $L(\alpha, p, k) \subset L(k/\alpha_1^{1/p})$ and lemma 9.4, we get the right-hand side of inequality (9.13). To prove the left-hand side of (9.13), observe that, in view of (5.2), $L(l) \subset L(\alpha, p, k)$. Thus, $\varphi_C(\alpha, p, k) \ge \varphi_C(l)$ and from lemma 9.4, we finally obtain

$$\varphi_C(\alpha, p, k) \ge \varphi_C(l) \ge \varphi'_C(1) \left(1 - \frac{1}{l} \right).$$

$\square$

Corollary 9.1 *For any $p \ge 1$, $k > 1$, and $\alpha = (\alpha_1, \alpha_2)$, we have*

$$\psi'_X(1) \left(1 - \frac{\left(\alpha_1 + \sqrt{\alpha_1^2 + 4\alpha_2 k^p} \right)^{1/p}}{2^{1/p} k} \right) \le \psi_X(\alpha, p, k) \le 1 - \frac{\alpha_1^{1/p}}{k},$$

and in the case of extremal space, we obtain

$$1 - \frac{\left(\alpha_1 + \sqrt{\alpha_1^2 + 4\alpha_2 k^p} \right)^{\frac{1}{p}}}{2^{1/p} k} \le \psi_X(\alpha, p, k) \le 1 - \frac{\alpha_1^{1/p}}{k}. \qquad (9.15)$$

For classes $L(\alpha, p, k)$, the theorem 9.16 takes the form:

Theorem 9.20 *For any $\alpha = (\alpha_1, \ldots, \alpha_n)$, $p \geq 1$ and $k > 1$, we have*

$$\varphi_C(\alpha, p, k) \geq \frac{\varphi'_C(1)}{pw'(k^p)}\left(1 - \frac{1}{k^p}\right),$$

$$\psi_X(\alpha, p, k) \geq \frac{\psi'_X(1)}{pw'(k^p)}\left(1 - \frac{1}{k^p}\right),$$

and

$$\varphi_X(\alpha, p, k) \geq \frac{\varphi'_X(1)}{pw'(k^p)}\left(1 - \frac{1}{k^p}\right).$$

Proof. Using (5.2), we get $L(l) \subset L(\alpha, p, k)$, where l is the solution of the equation (9.14). Thus, in view of lemma 9.4, for every $s < l$, we have

$$\varphi_C(\alpha, p, k) \geq \varphi_C(l) \geq \frac{s\varphi_C(s)}{s-1} \cdot \frac{l-1}{w(l^p)-1} \cdot \frac{k^p}{l} \cdot \frac{w(l^p)-1}{k^p}.$$

Hence, using lemma 9.4, we obtain

$$\varphi_C(\alpha, p, k) \geq \varphi'_C(1) \cdot \frac{l-1}{w(l^p)-1} \cdot \frac{k^p}{l} \cdot \left(1 - \frac{1}{k^p}\right).$$

In view of Lagrange's Mean Value Theorem, there exists $\theta \in (1, l)$ such that

$$\frac{w(l^p)-1}{l-1} = p\theta^{p-1}w'(\theta^p) \leq pk^{p-1}w'(k^p).$$

Since $k > l$, we finally obtain

$$\varphi_C(\alpha, p, k) \geq \frac{\varphi'_C(1)}{pk^{p-1}w'(k^p)} \cdot \frac{k^p}{l} \cdot \left(1 - \frac{1}{k^p}\right) \geq \frac{\varphi'_C(1)}{pw'(k^p)}\left(1 - \frac{1}{k^p}\right).$$

$\square$

The following theorem is a generalization of theorem 9.18.

Theorem 9.21 *For any $p \geq 1$ and $\alpha = (\alpha_1, \ldots, \alpha_n)$, we have*

$$\lim_{k \to \infty} \psi_X(\alpha, p, k) = 1.$$

Proof. It is enough to observe that for any $\alpha = (\alpha_1, \ldots, \alpha_n)$, $p \geq 1$, and $k > 1$ we have

$$\psi_X(l) \leq \psi_X(\alpha, p, k) \leq \psi_X\left(\frac{k}{\alpha_1^{1/p}}\right)$$

with $l > 1$ satisfying the equation (9.14).

$\square$

The estimates presented in the theorem 9.19 are not sharp. Indeed, if the length n of multi-index tends to infinity, then it may happen that solutions $l = l_n$ of the equation (9.14) decreases to 1 and this implies that the left-hand side of (9.13) tends to 0. For example, if we define the sequence of multi-indexes $\{\alpha^{(n)}\}_{n \geq 2}$ by

$$\alpha^{(n)} = \left(\frac{1}{n}, \underbrace{0, \ldots, 0}_{n-2}, 1 - \frac{1}{n}\right),$$

then for any $n \geq 2$ and for arbitrarily fixed $k > 1$, the equation (9.14) takes the form

$$\frac{1}{n}l_n^p + \left(1 - \frac{1}{n}\right)l_n^{np} = k^p.$$

Since

$$k^p = \frac{1}{n}l_n^p + \left(1 - \frac{1}{n}\right)l_n^{np} > \left(1 - \frac{1}{n}\right)l_n^{np},$$

we conclude that

$$1 < l_n < \frac{k^{1/n}}{\left(1 - \frac{1}{n}\right)^{1/np}}.$$

Letting n tend to infinity we obtain $\lim_{n \to \infty} l_n = 1$.

On the other hand, we have already seen in example 3.6 that, using a lipschitzian retraction $R : B \to S$ of the closed unit ball B onto its boundary S, we can construct a mapping $T : B \to S$ by putting $T = -R$, which is uniformly lipschitzian with $k(T^n) = k(R)$ for $n \geq 1$ and consequently for any $p \geq 1$ we have (see (5.3))

$$T \in \bigcap_{\alpha \in \mathfrak{I}} L(\alpha, p, k(R)).$$

Since $d(T) \geq 2/k(R)$ and $\psi(\alpha, p, \cdot)$ is nondecreasing with respect to k, we conclude that, for each $\alpha = (\alpha_1, \ldots, \alpha_n)$, $p \geq 1$ and $k \geq k(R)$,

$$\psi(\alpha, p, k) \geq \frac{2}{k(R)} > 0. \tag{9.16}$$

Consequently, for sufficiently large k the function $\psi(\cdot, \cdot, k)$ is separated from zero!

It is not clear whether the function $\psi(\alpha, p, \cdot)$ is continuous with respect to k. Hence, if we want to use the constant $k_0(X)$, then the estimate (9.16) takes the form: for any $p \geq 1$, $\alpha = (\alpha_1, \ldots, \alpha_n)$, and $k > k_0(X)$,

$$\psi(\alpha, p, k) \geq \frac{2}{k_0(X)} > 0.$$

Below we list some estimates obtained via method described above.

- Since for extremal and cut invariant subspace X of $B(K)$ we have $k_0(X) \leq 4(2+\sqrt{3})$, for any $p \geq 1$, $\alpha = (\alpha_1, \ldots, \alpha_n)$ and $k > 4(2+\sqrt{3})$,

$$\psi_X(\alpha, p, k) \geq \frac{1}{2(2+\sqrt{3})}.$$

For the spaces H, $BC(\mathbb{R})$, $C[0,1]$, c, c_0, ℓ_1, $L_1(0,1)$, and $BC_z(M)$, the upper bound for $k_0(X)$ has been obtained by constructing a retraction satisfying the Lipschitz condition with the Lipschitz constant, which is equal to the presented upper bound. Hence, for any $p \geq 1$ and $\alpha = (\alpha_1, \ldots, \alpha_n)$, we can write:

- $\psi_H(\alpha, p, 28.99) \geq 0.068989$;

- $\psi_{BC(\mathbb{R})}(\alpha, p, 4(2+\sqrt{3})) \geq 1 - \frac{\sqrt{3}}{2}$;

- $\psi_{C[0,1]}(\alpha, p, 4(2+\sqrt{3})) \geq 1 - \frac{\sqrt{3}}{2}$;

- $\psi_c(\alpha, p, 4(2+\sqrt{3})) \geq 1 - \frac{\sqrt{3}}{2}$;

- $\psi_{c_0}(\alpha, p, 4(2+\sqrt{3})) \geq 1 - \frac{\sqrt{3}}{2}$;

- $\psi_{\ell_1}(\alpha, p, 8) \geq \frac{1}{4}$;

- $\psi_{L_1(0,1)}(\alpha, p, 8) \geq \frac{1}{4}$;

- $\psi_{BC_z(M)}(\alpha, p, 2(2+\sqrt{2})) \geq 1 - \frac{\sqrt{2}}{2}$.

Note that we do not claim that for any multi-index α and $p \geq 1$ we have $\varphi_X(\alpha, p, 1) = 0$. However, some particular cases has been solved in previous chapter. Below, we list known facts in the language of function φ_X:

Conclusion 9.1 *In view of theorem 8.1, for $p \geq 1$ and a multi-index $\alpha = (\alpha_1, \alpha_2)$ with $\alpha_2^p + \alpha_2 \leq 1$, we have $\varphi_X(\alpha, p, 1) = 0$.*

Conclusion 9.2 *In view of theorem 8.3, for any $p \geq 1$ and $\alpha = (\alpha_1, \ldots, \alpha_n)$ such that*

$$(1 - \alpha_1)(1 - \alpha_1^{\frac{n-1}{p}}) \leq \alpha_1^{\frac{n-1}{p}}(1 - \alpha_1^{\frac{1}{p}}),$$

we have $\varphi_X(\alpha, p, 1) = 0$.

Conclusion 9.3 *In view of theorem 8.23, for any $\alpha = (\alpha_1, \alpha_2, \alpha_3)$ with $\alpha_1 \geq \frac{1}{2}$ and $\alpha_2 \geq \frac{1}{2} - \alpha_1^2$, we have $\varphi_X(\alpha, 1, 1) = 0$.*

Conclusion 9.4 *In view of theorem 8.11, if X is a Banach space with $\epsilon_0(X) < 1$, then for $p \geq 1$ and multi-index $\alpha = (\alpha_1, \ldots, \alpha_n)$ such that*

$$\left(\sum_{j=1}^{n} \left(\sum_{i=j}^{n} \alpha_i \right) \right)^{1/p} < \kappa(X),$$

we have $\varphi_X(\alpha, p, 1) = 0$.

Conclusion 9.5 *In view of theorem 8.15, for every Hilbert space H and multi-index $\alpha = (\alpha_1, \ldots, \alpha_n)$ such that*

$$\left(\sum_{j=1}^{n} \left(\sum_{i=j}^{n} \alpha_i \right) \right) < 2,$$

we have $\varphi_H(\alpha, 2, 1) = 0$.

Conclusion 9.6 *In view of theorem 8.17, if X is a Banach space with $\epsilon_0(X) < 1$, then for each multi-index $\alpha = (\alpha_1, \ldots, \alpha_n)$ of length $n \geq 2$ and $p = \frac{\ln(n)}{\ln(\kappa(X))}$, we have $\varphi_X(\alpha, p, 1) = 0$.*

Conclusion 9.7 *In view of theorem 8.18, for every Hilbert space H and for every multi-index $\alpha = (\alpha_1, \ldots, \alpha_n)$ of length $n \geq 2$ and $p = \frac{2\ln(n)}{\ln(2)}$, we have $\varphi_H(\alpha, p, 1) = 0$.*

Conclusion 9.8 *In view of theorem 8.16, for every Hilbert space H and for every multi-index $\alpha = (\alpha_1, \alpha_2)$, we have $\varphi_H(\alpha, 2, 1) = 0$.*

Bibliography

[1] D. Alspach. A fixed point free nonexpansive map. *Proc. Amer. Math. Soc.*, 82:423–424, 1981.

[2] M. Annoni and E. Casini. An upper bound for the Lipschitz retraction constant in l_1. *Studia Math.*, 180:73–76, 2007.

[3] A. D. Arvanitakis. A proof of the generalized Banach Contraction Conjecture. *Proc. Amer. Math. Soc.*, 131(12):3647–3656, 2003.

[4] J. B. Baillon. Quelques aspects de la théorie des points fixes dans les éspaces Banach. *Seminaire d'Analysis Fonctionnelle, Ecole Polytechnique I*, 1978–79.

[5] E. J. Barbeau. *Polynomials*. Springer-Verlag New York Inc., 1989.

[6] M. Baronti, E. Casini, and C. Franchetti. The retraction constant in some Banach spaces. *J. Approx. Theory*, 120:296–308, 2003.

[7] T. Domínguez Benavides. Normal structure coefficient in l_p spaces. *Proc. Royal Soc. Edinburgh*, 117(-A):299–303, 1991.

[8] T. Domínguez Benavides. Fixed point theorems for uniformly lipschitzian mappings and asymptotically regular mappings. *Nonlinear Anal.*, 32:15–27, 1998.

[9] T. Domínguez Benavides. A renorming of some nonseparable Banach spaces with the Fixed Point Property. *J. Math. Anal. Appl.*, 350(2):525–530, 2009.

[10] Y. Benyamini and Y. Sternfeld. Spheres in infinite dimensional normed spaces are Lipschitz contractible. *Proc. Amer. Math. Soc.*, 88:439–445, 1983.

[11] K. Bolibok. Constructions of lipschitzian mappings with non zero minimal displacement in spaces $L^1(0,1)$ and $L^2(0,1)$. *Ann. Univ. Mariae Curie-Sklodowska Sec. A*, 50:25–31, 1996.

[12] K. Bolibok. *Minimal displacement and retraction problems for balls in Banach spaces*. Maria Curie-Sklodowska University, (Ph.D. Thesis), 1999 (in Polish).

[13] K. Bolibok. Minimal displacement and retraction problems in infinite-dimensional Hilbert spaces. *Proc. Amer. Math. Soc.*, 132(4):1103–111, 2004.

[14] K. Bolibok. The minimal displacement problem in the space l^∞. *Cent. Eur. J. Math.*, 10(6):2211–2214, 2012.

[15] M. S. Brodskii and D. P. Milman. On the center of a convex set. *Dokl. Akad. Nauk SSSR*, 59:837–840, 1948 (in Russian).

[16] L. E. J. Brouwer. Über Abbildungen von Mannigfaltigkeiten. *Math. Ann.*, 71:97–115, 1912.

[17] F. E. Browder. Nonexpansive nonlinear operators in a Banach space. *Proceedings of the National Academy of Sciences of the U.S.A.*, 54:1041–1043, 1965.

[18] W. L. Bynum. Normal structure coefficients for Banach spaces. *Pacific J. Math.*, 86:427–436, 1980.

[19] E. Casini and E. Maluta. Fixed points for uniformly lipschitzian mappings in spaces with uniformly normal structure. *Nonlinear Anal.*, 9:103–108, 1985.

[20] J. A. Clarkson. Uniformly Convex Spaces. *Trans. Amer. Math. Soc.*, 40:396–414, 1936.

[21] D. G. deFigueiredo and L. A. Karlovitz. On the radial projection in normed spaces. *Bull. Amer. Math. Soc.*, 73:364–368, 1967.

[22] P. N. Dowling, W. B. Johnson, C. J. Lennard, and B. Turett. The optimality of James' Distortion Theorems. *Proc. Amer. Math. Soc.*, 125:167–174, 1997.

[23] P. N. Dowling and C. J. Lennard. Every nonreflexive subspace of $l_1[0,1]$ fails the fixed point property. *Proc. Amer. Math. Soc.*, 125:443–446, 1997.

[24] P. N. Dowling, C. J. Lennard, and B. Turett. Some fixed point results in l_1 and c_0. *Nonlinear Anal.*, 39:929–936, 2000.

[25] D. J. Downing and B. Turett. Some properties of the characteristic of convexity relating to fixed point theory. *Pacific J. Math.*, pages 342–350, 1983.

[26] M. Edelstein. The construction of an asymptotic center with a fixed-point property. *Bull. Amer. Math. Soc.*, 78:206–208, 1972.

[27] C. Franchetti. On the radial constant of real normed spaces. *Cienc. Tecn.*, 3:1–9, 1979.

[28] Jesús García-Falset, Enrique Llorens-Fuster, and Eva M. Mazcuñán-Navarro. Uniformly nonsquare Banach spaces have the fixed point property for nonexpansive mappings. *Journal of Functional Analysis*, 233:494–514, 2005.

[29] K. Goebel. On the minimal displacement of points under lipschitzian mappings. *Pacific J. Math.*, 48:151–163, 1973.

[30] K. Goebel. On the structure of minimal invariant sets for nonexpansive mappings. *Ann. Univ. Mariae Curie-Skłodowska*, 29:73–77, 1975.

[31] K. Goebel. *Concise Course on Fixed Point Theorems*. Yokohama Publishers, 2002.

[32] K. Goebel and E. Casini. Why and how much the Brouwer's Fixed Point Theorem fails in noncompact setting? *Milan. J. Math.*, 78(2):371–394, 2010.

[33] K. Goebel and M. A. Japón-Pineda. On a type of generalized nonexpansiveness. *Proceedings of the 8th International Conference on Fixed Point Theory and its Applications, Chiang-Mai*, pages 71–82, 2007.

[34] K. Goebel and W. Kaczor. Remarks on failure of Schauder's theorem in noncompact settings. *Ann. Univ. Mariae Curie Sklodowska Sect. A*, 51:99–108, 1997.

[35] K. Goebel and W. A. Kirk. A fixed point theorem for transformations whose iterates have uniform Lipschitz constant. *Studia Math.*, 67:135–140, 1973.

[36] K. Goebel and W. A. Kirk. *Topics in metric fixed point theory*. Cambridge University Press, 1990.

[37] K. Goebel and T. Kuczumow. Irregular convex sets with the fixed point property for nonexpansive mappings. *Colloq. Math.*, 40:259–264, 1978.

[38] K. Goebel and G. Marino. A note on minimal displacement and optimal retraction problems. *Proceedings of the Seventh International Conference on Fixed Point Theory and its Applications, Guanajuato, Mexico, July 17–23, 2005*, Yokohama Publishers, 2006.

[39] K. Goebel, G. Marino, L. Muglia, and R. Volpe. The retraction constant and minimal displacement characteristic of some Banach spaces. *Nonlinear Anal.*, 67:735–744, 2007.

[40] K. Goebel and Ł. Piasecki. A new estimate for the optimal retraction constant. *Proceedings of the Second International Symposium on Banach and Function Spaces 2006, Kitakyushu, Japan, September 14-17, 2006*, pages 77–83, Yokohama Publishers, 2008.

[41] K. Goebel and Ł. Piasecki. Minimal displacement and related problems, revisited. *Proceedings of the Fourth International Symposium on Banach and Function Spaces 2012, Kitakyushu, Japan, September 12–15, 2012,* Yokohama Publishers, 2014.

[42] K. Goebel and B. Sims. Mean Lipschitzian Mappings. *Contemp. Math.,* 513:157–167, 2010.

[43] D. Göhde. Zum Prinzip der kontraktiven Abbildung. *Math. Nachr.,* 30:251–258, 1965.

[44] H. K. Hsiao and R. Smarzewski. On best Lipschitz constants of radial projections. *Atti Sem. Mat. Fis. Univ. Modena,* 43:511–517, 1995.

[45] S. Kakutani. Topological properties of the unit sphere of a Hilbert space. *Proc. Imp. Acad. Tokyo,* 19:269–271, 1943.

[46] L. A. Karlovitz. Existence of fixed points for noncxpansive mappings in spaces without normal structure. *Pacific J. Math.,* 66:153–156, 1976.

[47] L. A. Karlovitz. On nonexpansive mappings. *Proc. Amer. Math. Soc.,* 55:153–159, 1976.

[48] W. A. Kirk. A fixed point theorem for mappings which do not increase distances. *Amer. Math. Monthly,* 72:1004–1006, 1965.

[49] W. A. Kirk and B. Sims (editors). *Handbook of metric fixed point theory.* Kluwer Academic Publishers, 2001.

[50] M. D. Kirzbraun. Über die Zussamenziehende und Lipschistsche Transformationen. *Fund. Math.,* 22:77–108, 1934.

[51] V. Klee. Convex bodies and periodic homeomorphisms in Hilbert spaces. *Trans. Amer. Math. Soc.,* 74:10–43, 1953.

[52] V. Klee. Some topological properties of convex sets. *Trans. Amer. Math. Soc.,* 78:30–45, 1955.

[53] T. Komorowski and J. Wośko. A remark on the retracting of a ball onto a sphere in an infinite dimensional Hilbert space. *Math. Scand.,* 67:223–226, 1990.

[54] E. A. Lifshitz. Fixed point theorems for operators in strongly convex spaces. *Voronez Gos. Univ. Trudy Mat. Fak. (in Russian),* 16:23–28, 1975.

[55] P. K. Lin. Stability of the fixed point property of Hilbert space. *Proc. Amer. Math. Soc.,* 127:3573–3581, 1999.

[56] P. K. Lin. There is an equivalent norm on ℓ_1 that has the fixed point property. *Nonlinear Anal.,* 68(8):2303–2308, 2008.

[57] P. K. Lin and Y. Sternfeld. Convex sets with the Lipschitz fixed point property are compact. *Proc. Amer. Math. Soc.*, 93:633–639, 1985.

[58] E. Maluta. Uniformly normal structure and related coefficients. *Pacific J. Math.*, 111:357–369, 1984.

[59] D. Mauldin. *The Scottish Book: mathematical problems from the Scottish Cafe*. Birkhäuser, Boston, 1981.

[60] B. Maurey. Points fixes des contractions de certain faiblement compacts de l^1. *Seminaire d'Analyse Fonctionelle, Exposé VIII, École Polytechnique, Centre de Mathematiques*, 1980–1981.

[61] E. Mazcuñán-Navarro. Stability of the fixed point property in Hilbert spaces. *Proc. Amer. Math. Soc.*, 134(1):129–138, 2006.

[62] J. Merryfield and J. D. Stein Jr. A generalization of the Banach Contraction Principle. *J. Math. Anal. Appl.*, 273:112–120, 2002.

[63] B. Nowak. On the Lipschitz retraction of the unit ball in infinite dimensional Banach space onto boundary. *Bull. Acad. Polon. Sci.*, 27:861–864, 1979.

[64] V. Pérez-García and Ł. Piasecki. Lipschitz constants for iterates of mean lipschitzian mappings. *Nonlinear Anal.*, 74:5643–5647, 2011.

[65] V. Pérez-García and Ł. Piasecki. On mean nonexpansive mappings and the Lifshitz constant. *J. Math. Anal. Appl.*, 396(2):448–454, 2012.

[66] V. Pérez-García and Ł. Piasecki. Spectral radius for mean lipschitzian mappings. *Proceedings of the 10th International Conference on Fixed Point Theory and Its Applications, July 9–18, 2012, Cluj-Napoca, Romania*, pages 209–216, House of the Book of Science 2013.

[67] Ł. Piasecki. *O retrakcjach kul na sfery*. Mariae Curie-Sklodowska University (M.Sc. Thesis), 2007 (in Polish).

[68] Ł. Piasecki. Retracting ball onto sphere in $BC_0(\mathbb{R})$. *Topol. Methods Nonl. An.*, 33:307–313, 2009.

[69] Ł. Piasecki. Retracting a ball onto a sphere in some Banach spaces. *Nonlinear Anal.*, 74:396–399, 2011.

[70] Ł. Piasecki. *Przekształcenia średnio lipschitzowskie*. Mariae Curie-Sklodowska University (Ph.D. Thesis), 2012 (in Polish).

[71] S. Prus. On Bynum's fixed point theorem. *Atti Sem. Mat. Fis. Univ. Modena*, 38:535–545, 1990.

[72] S. Prus. Banach spaces which are uniformly noncreasy. *Nonlinear Anal.*, 30:2317–2324, 1997.

[73] J. J. Schaffer. *Geometry of spheres in normed spaces.* Marcel Dekker, Lecture Notes in Pure and Applied Mathematics, 1976.

[74] J. Schauder. Der Fixpunktzats in Funktionalraumen. *Studia Math.*, 2:171–180, 1930.

[75] R. L. Thele. Some results on the radial projection in Banach spaces. *Proc. Amer. Math. Soc.*, 42:483–486, 1974.

[76] F. A. Valentine. A Lipschitz condition preserving extension for a vector function. *Amer. J. Math.*, 67:83–93, 1945.

[77] V. Zizler. On some rotundity and smoothness properties of Banach spaces. *Dissertationes Math.*, 87, 1971.

Index

Milton Keynes UK
Ingram Content Group UK Ltd.
UKHW040102071024
449327UK00019B/749